Introduction to Plasma Physics

Series Editor
Laurence Rezeau

Introduction to Plasma Physics

Gérard Belmont
Laurence Rezeau
Caterina Riconda
Arnaud Zaslavsky

ELSEVIER

First published 2019 in Great Britain and the United States by ISTE Press Ltd and Elsevier Ltd

ISTE Press Ltd
27-37 St George's Road
London SW19 4EU
UK

www.iste.co.uk

Elsevier Ltd
The Boulevard, Langford Lane
Kidlington, Oxford, OX5 1GB
UK

www.elsevier.com

Notices

For information on all our publications visit our website at http://store.elsevier.com/

British Library Cataloguing-in-Publication Data
A CIP record for this book is available from the British Library
Library of Congress Cataloging in Publication Data
A catalog record for this book is available from the Library of Congress
ISBN 978-1-78548-306-6

Printed and bound in the UK and US

Contents

Introduction

It can be said that the "plasma" state constitutes the "fourth state of matter". It is less well-known than the other three states of matter (solid, liquid and gas) because it is less present in our immediate environment. However, it must be known that neutral matter, which constitutes the largest part of this environment, is an exception in the universe. In most of the latter, matter consists either totally or partially of charged particles (particularly, electrons and protons) which are free and not bound within neutral atoms and molecules; these charged particle gases are called "plasmas". Their main property, which distinguishes them from neutral gases, is that they closely interact with the electromagnetic field, on the one hand, because the movement of particles is governed by fields, and on the other hand, because the ensemble of particles is itself a source of fields by the charge density and the currents that these movements cause.

Plasma physics is thus at the intersection of statistical physics and electromagnetism:

– for the physics of natural plasmas, its most developed fields of application are external geophysics (ionosphere/magnetosphere of the Earth and other planets, aurora borealis, etc.), solar and stellar physics (solar corona, solar wind, etc.) and astrophysics (galactic jets, etc.);

– for laboratory plasmas, they play a very important role in studies concerning nuclear fusion, either by magnetic confinement (tokamaks) or by inertial confinement (laser fusion) and in the production of energy particles via plasma accelerator;

– partially ionized gases also constitute the state of matter encountered in discharges (lightning, neon tubes, etc.) as well as in many technological applications (surface treatment, deposition, etching, etc.).

1

What Is Plasma?

The plasmas, which will be presented in this chapter, resemble gases, but because they are constituted of free charged particles, the physics that govern their dynamics is radically different. First, the charged particles' motion is determined by electromagnetic fields, and second, the fields are created by charge and current densities caused by these particles. This coupling will be illustrated in a simple example, called "plasma oscillation". In this fundamental example, we will see how all field fluctuations are accompanied by matter movements and, vice versa, how every matter movement is accompanied by a field fluctuation.

1.1. Under what conditions is matter in the plasma state?

The term "plasma" was introduced for the first time by I. Langmuir in 1928 when he studied the ionized gas behavior in discharge tubes because the ion oscillations observed in these tubes were reminiscent of the oscillations observed in a gelatinous environment (plasma means gelatinous matter, or matter that can be modeled, in Greek). Plasma then appears as a "fourth" state of matter, a gaseous and ionized medium in which particle dynamics is dominated by electromagnetic forces: other forces, such as gravity, are often negligible in this kind of system. Neutral media are, in fact, also made up of electrons and protons, which are charged particles; however, in this type of medium, they are bound within globally neutral atoms and molecules. What distinguishes plasma from neutral media is the presence of "free" charged particles, that is, particles that can move independently (in opposite directions according to the sign of their charge), thus creating currents and deviations from neutrality. The term "plasma" may in fact be extended to all media (equivalents of gases, as well as liquids and solids) in which there are such free charged particles. We will limit ourselves in this book essentially to plasmas, which are the equivalents of gases and which must be named, more precisely, "weakly correlated plasmas". We will therefore consider media that are sufficiently tenuous.

The development of plasma physics followed these first discoveries, in conjunction with research on radio-communications. As early as 1901, G. Marconi observed the reflection of the waves on what he thought was the atmosphere, but it was in fact the ionosphere. The idea that our atmosphere is ionized from a certain altitude was brought up by E. Appleton in 1925; he thus launched the study of natural plasmas, which has gradually become that of astrophysical plasmas. In the laboratory, studies have continued beyond discharges, particularly with research on electron beams as sources of coherent radiation (klystrons), and they have progressed to a much more intensive stage with the initiation of research on controlled nuclear fusion, circa 1955. More recently, work has been undertaken to study the interactions between plasmas and surfaces, leading to surface treatments in mechanics or microelectronics thanks to plasmas. It has also been shown that laser-created plasmas can behave as sources of rapid particles or radiation, that is, as miniature accelerators, which offer an alternative to traditional accelerators. Plasma research is therefore very active in the fields of astrophysics, fusion and industrial applications.

Under the "normal" temperature and pressure conditions in which we live, particles are naturally bound in the form of neutral atoms and molecules (although, in the rest of the universe, this "fourth" state of matter is the normal state). To create plasma from such a neutral gas, it is necessary to provide energy to remove one or more electrons from each atom. It is therefore necessary that sufficient energy be provided to the atoms so that they are partially, or even totally, ionized. This energy can be provided in many ways.

1.1.1. *Electric discharges*

As I. Langmuir first showed, a gas can be ionized and a plasma created when we produce an electric discharge in it. Such plasmas are indeed present in our familiar surroundings, for instance, in certain types of lamps, neon signs and lightning (see Figure 1.1). Others are also created in the same way for research purposes, as will be seen in the following (see Figure 1.2).

Figure 1.1. *Lightning (source: Y. Faust). For a color version of this figure, see www.iste.co.uk/belmont/plasma.zip*

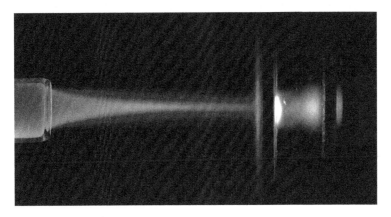

Figure 1.2. *Discharge in a helium jet at atmospheric pressure. The discharge propagates to the surface of a glass cell in which a second discharge in helium at low pressure is initiated (source: O. Guaitella). For a color version of this figure, see www.iste.co.uk/belmont/plasma.zip*

1.1.2. *Heating*

In nature, ionization is often produced by thermal collisions between atoms (or molecules) if their temperature is high enough. For a set of atoms maintained in local thermodynamic equilibrium by collisions, when the atoms' average kinetic energy $k_B T$ becomes of the order of the atom's ionization energy ($\approx eV$), the fraction of ionized atoms becomes significant. The effect of temperature on the ionization state of a gas was first described by astrophysicist M. Saha in 1920, and is summarized with the equation called "Saha's ionization equation", which provides, in the hypothesis of a weakly ionized plasma in a thermodynamic equilibrium state, the atom density in each ionization state as a function of the ionization energies and the temperature. Figure 1.3 is a combination of images of the solar corona taken by the EIT telescope aboard the SOHO probe. These images show the presence of ionized chemical elements that can only exist at very high temperatures. From these observations, we therefore immediately deduce an order of magnitude of the temperature of the solar corona and the confirmation that the medium is indeed composed of plasma.

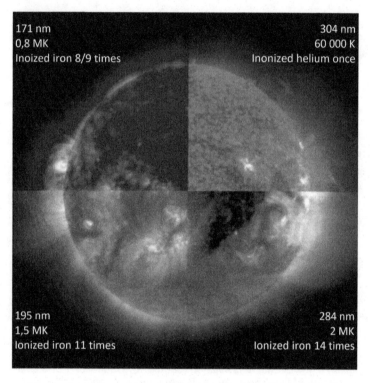

Figure 1.3. *Image of the solar corona realized in four ultraviolet wavelengths (source: SOHO/EIT consortium. SOHO is a collaborative ESA/NASA mission). For a color version of this figure, see www.iste.co.uk/belmont/plasma.zip*

NOTE.– Often in plasma physics, temperature is indicated by the energy associated with it in eV; the conversion factor is 1 eV = 11,605 K ≈ 10^4 K.

1.1.3. *Radiation absorption*

A third way to ionize atoms may be via radiation absorption. The absorption of a photon by an atom can produce an ion and an electron. If the recombination is slow enough, plasma is formed. An example of plasma created in this way at low temperature is that of the Earth's ionosphere or that of any other planet with an atmosphere (Figure 1.4). Solar ultraviolet radiation is absorbed by the atmosphere's upper layers and ionizes the atoms and molecules these layers are made of. The terrestrial ionosphere has a maximum electron density of around 300 km. It is larger during the day, when the atmosphere is hit by the Sun's radiation, which confirms the plasma's creation mechanism. The electrons present on the night side were created on the day side and are driven by the atmosphere's rotation.

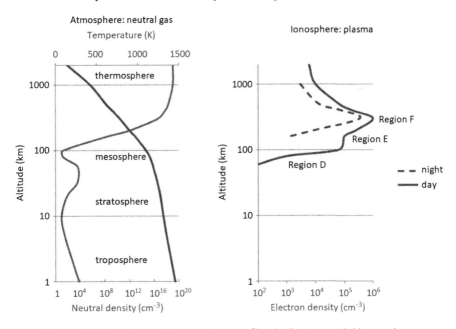

Figure 1.4. *Density and temperature profiles in the terrestrial ionosphere. For a color version of this figure, see www.iste.co.uk/belmont/plasma.zip*

1.1.4. *Different types of plasmas*

The plasma state therefore groups very different media. In the natural state, plasmas are present everywhere in the universe, including at less than a hundred kilometers above our heads. The ionosphere is an example of partially ionized plasma, where atoms and neutral molecules coexist with electrons and ions, and give rise to a large number of chemical reactions. At a higher altitude (magnetosphere), the ionization becomes total. Many other astrophysical plasmas are also completely ionized.

Artificial plasmas can also be produced, with a wide range of accessible physical parameters, depending, for example, on whether one considers discharge plasmas or fusion plasmas (the temperature must be very high to reach fusion).

We will also distinguish between magnetized and non-magnetized plasmas. The external magnetic field in plasma is often introduced, by means of coils or magnets, in order to keep it confined in an experimental device. Charged particles move along the magnetic field lines in their gyration movement (see Chapter 2); it is then possible to impose magnetic field configurations such that the particles do not escape (or such that very little escape) from the confinement device. Such plasma "contains" a magnetic field constraining the movement of the particles, and it is said to be magnetized. In nature, too, we find magnetized plasmas, either by the magnetic field associated with a star, a planet or the interstellar medium, or by a magnetic field auto-generated by the plasma itself. In the absence of such a field, we will refer to non-magnetized plasmas.

Hot or cold plasma? The literature often assigns the qualifiers "hot" or "cold" to plasma. These terms are, however, imperfectly defined, and may cover different notions. Most of the plasmas created in the laboratory (by discharge in particular) can be called cold plasmas, in which the ionization rates are low and the temperature of the ions is in equilibrium via collisions with that of the neutrals, and are thus close to the ambient temperature ~273 K. On the contrary, the electrons of these plasmas have much higher temperatures (of the order of at least a few electron volts) and undergo few collisions with ions and neutrals. Their dynamics is then mainly governed by the electromagnetic field. In this sense, hot plasma is, in contrast, highly ionized plasma in which the temperatures of ions, as those of electrons, can be much higher than the ambient temperature. This definition is, of course, very empirical and leaves a fuzzy notion about how to name certain environments with "intermediate" behaviors. Another, more theoretical way of attributing the qualifiers "hot" or "cold" to plasma (or, more precisely, to a plasma population), is to look at the importance of the pressure gradient in the balance of the voluminal forces acting on the plasma. This definition will be used systematically in this book. It will be said that the plasma is cold when the pressure term is negligible compared to the other forces involved (the electric field, for example) and hot when, on the contrary, this pressure term cannot be neglected.

Figure 1.5 groups together the orders of magnitude characteristic of some plasmas in order to fix ideas on what "plasma" can be. For comparison, we recall that the atmosphere in which we live has a density (of neutrals) of about 3×10^{25} m^{-3} for a temperature of 273 K. Thus, we can see that the conditions that we will discover by exploring plasmas can be radically different from those we are familiar with. Note that, in this figure, the flame and the metal are present although they are not really plasmas. The flame is so weakly ionized that its behavior is similar to that of a gas. The electron gas in a metal has a quantum behavior, which also makes it a very special case.

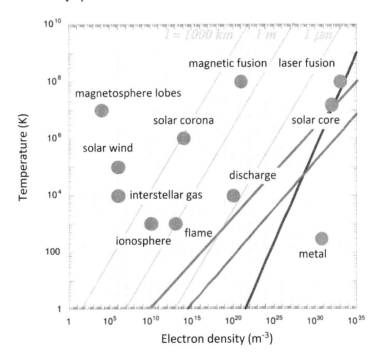

Figure 1.5. *Order of magnitude of temperatures and densities of some plasmas. The logarithmic scales show the extent of the range of parameters found in nature and in laboratory plasmas. The lines in bold represent the plasmas for which the inter-particular distance d is equal to the Landau length r_0 (blue), the Debye length λ_{De} (yellow) or the de Broglie wavelength λ_{dB} (red). The green lines represent the value of the average free path of the electrons (see section 1.2 for the discussion on these lines). For a color version of this figure, see www.iste.co.uk/belmont/plasma.zip*

1.2. Plasma diagnostics: remote or *in situ*

The observation of plasmas can be done indirectly, such as, for example, when Marconi discovered the ionosphere. It is the effect of plasma on wave propagation that shows the existence of this plasma. Indeed, like most gases, plasmas are not visible. Visible plasmas are those that emit light via an excitation–de-excitation mechanism of their neutrals and ions. As an example, we can refer to auroras, with their amazing colors.

The aurora in Figure 1.6 is associated with a ray of atomic oxygen. At about 100 km of altitude, the composition of the atmosphere is very different from what it is on the ground and atomic oxygen is one of the most abundant elements. The electrons that create the aurora are precipitated along the magnetic field lines (see Chapter 2) and they excite the atmosphere's molecules, which are then de-excited by emitting light. The colors of the auroras depend on the altitude and reflect the atomic composition.

Figure 1.6. *Green aurora (source: F. Mottez, Tromsø, Norway, 2013).*
For a color version of this figure, see www.iste.co.uk/belmont/plasma.zip

However, not all radiation is visible and some plasmas emit in wavelengths that are invisible to the eye but are accessible thanks to measurements by spectroscopy.

Figure 1.3 was obtained from non-visible radiation emitted by the Sun, converted into false colors. Among these radiations, we find emission lines of iron: the figure thus demonstrates the presence of iron in the corona. Spectroscopy thus provides access to the composition, density and temperature conditions of the plasma. It actually allows us to obtain much more information: for example, the Zeeman effect can be used to measure the magnetic field in the plasma by splitting the emission lines (Figure 1.7). The polarization of the transmitted waves can also be used to obtain the direction of the magnetic field in the plasma.

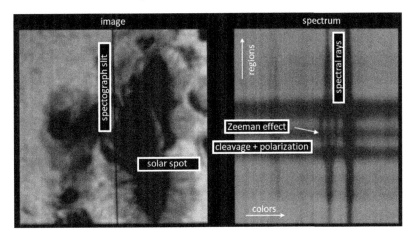

Figure 1.7. *On the left is the image of a sunspot. The black line is the slit of a spectrograph, which makes it possible to draw the spectrum on the right. Two lines cross the entire image. In the central part (which corresponds to the spot), we see that the line on the left-hand side separates and that the one on the right-hand side widens. It is therefore in the spots that the field is the most intense (source: J.F. Donati). For a color version of this figure, see www.iste.co.uk/belmont/plasma.zip*

Spectroscopy is a means of obtaining information on the plasma by observing it at a distance. Measurements can also be made in the plasma itself (referred to as "*in situ*" measurements). The oldest sensor is probably the Langmuir probe. It is a small electrode that is introduced into the plasma and whose potential is varied. The current–voltage characteristic of the probe makes it possible to obtain information on the plasma's density and temperature. This instrument is commonly used both in laboratory plasmas and on space missions that explore natural plasmas in the solar wind and planetary environments. Onboard these missions are also instruments that measure electrical and magnetic fields (Figure 1.8), both continuous (DC) and alternating (AC). These instruments are not specific to plasmas but are necessary to know all the parameters that intervene in the physics of the medium. There are also particle detectors that are more sophisticated than the Langmuir probe that provide access to the plasma's composition (see Chapter 3).

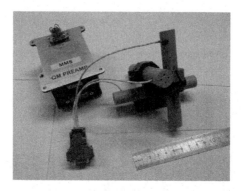

Figure 1.8. *AC magnetic field sensor with flux feedback. Qualification model (QM) manufactured at the LPP for the MMS (NASA) mission. (Right) three antennas; (left) the preamplifier box. For a color version of this figure, see www.iste.co.uk/belmont/plasma.zip*

This *in situ* observation of plasmas is inseparable from the progress of space exploration. The first measurements were made in the 1960s by instruments carried on rockets (see Figure 1.9). They were followed by satellite missions around the Earth, and then around the other planets of the solar system.

Figure 1.9. *Dragon rocket (LPP collection). Several versions of this series were launched during the austral summer 1966–1967 from the Dumont d'Urville base by the CNES. The instruments are at the extremities of the arms that are deployed after launching. For a color version of this figure, see www.iste.co.uk/belmont/plasma.zip*

This space exploration continues and several examples of measurements will be used throughout this book.

1.3. Effects that dominate physics

In a neutral gas, collisions and external forces such as gravity govern the overall dynamics of the fluid. Under the usual conditions of temperature and pressure (those that we call "normal"), these collisions are so numerous that we hardly deviate from equilibrium. These collisions also ensure diffusion and transport when the system is out of equilibrium. The collisions we are talking about in this case are collisions between neutral particles; they are therefore governed by short-range forces that can be modeled, without making too much of an oversimplified approximation by the so-called "billiard balls" model.

In totally ionized plasma, the interactions that occur between particles are electromagnetic and, therefore, fundamentally different. Unlike interactions between neutral particles, these interactions are of long range, the field created by a charge only decreasing by a factor of $1/r^2$. A given particle may be sensitive to a very close neighbor ("close binary interaction"), but it is a negligible effect in weakly correlated plasmas; it is especially sensitive to all the others, much more distant, via the electromagnetic fields that they create. On a certain scale (very small but sufficient to have a statistic relating to a large number of particles), we can distinguish a completely collective "mean" field and a field that varies randomly, described as "thermal noise". This random part is responsible for what we call "collisions" in these plasmas. This point will be specified in Chapter 3. In the case of partially ionized plasma, which we will not deal with here, the two types of collisions (with neutrals and between charged particles) intervene in the physics of the medium.

When we more precisely quantify the relative effects of the mean field and the thermal noise (Chapter 3), we see a very important characteristic scale of the plasma: the Debye length. Concerning the mean field, we will now show that this scale manifests itself in a simple way when a charged body is placed in plasma: a "sheath" is formed, which "screens" the electric field created by this charged body, that is, beyond which this field tends towards zero. This sheath's thickness is the Debye length λ_D. This length is important because it characterizes the plasma's tendency to maintain quasi-neutrality at scales larger than λ_D.

1.3.1. Screening phenomenon

Consider an isolated sphere charged with a charge q. Its electrostatic potential is given by:

$$V = \frac{q}{4\pi\varepsilon_0 r}$$

The same sphere in plasma will tend to attract particles of the opposite sign and repel particles of the same sign until equilibrium, where the thermal movement of the particles compensates the force of the electric field. The result is that the potential of the charged sphere in the plasma is given by:

$$V = \frac{q}{4\pi\varepsilon_0 r} e^{-r/\lambda_D}$$

(This so-called Debye–Hückel potential was initially established in chemistry to characterize the sheath occurring around an electrode in an electrolyte.)

In plasma, in relation to the mass of electrons, which are much lower than that of ions, we can consider that screening is essentially due to electrons (although, in fact, ions also participate). We therefore define the Debye electronic length $\lambda_{De} = \sqrt{\frac{\varepsilon_0 k_B T}{ne^2}}$, which is the reference screening scale in the plasma. Another example of screening is shown in Figure 1.10, in which the effect of plasma screening around charged conductive wires is shown. In the center of the figure are the conductive wires, around which there is a sheath of width λ_D. On a smaller scale or in the same order as the Debye length, we can clearly see a charge accumulation of sign opposite to that of the charged wires, but at a greater distance, the plasma is neutral. This illustrates the fact that, in plasma, the effect of a deviation from neutrality can only have a limited range in the steady state.

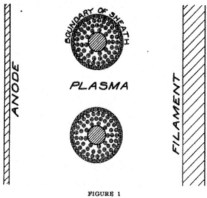

FIGURE 1

Positive ion sheaths around grid wires in a thermionic tube containing gas.

Figure 1.10. *Debye screening around the wires. The sheath boundaries are explicitly shown with a globally neutral plasma around it. This is the first figure with the word plasma, published in an article by Hull and Langmuir (1928)*

1.3.2. *Binary interactions or collective forces?*

The role of collisions between neutral and charged particles is described by the notion of the mean free path. The mean free path is the distance over which the trajectory of a test particle significantly deviates from its "ideal" trajectory, due to the only collective field. In a neutral gas, its meaning is simple: the particle trajectory is a broken line whose sections have an average size called the average free path. However, in a plasma, the trajectory is completely different and the mean free path estimation can be tricky due to interactions with distant particles caused by Coulomb force, which is a long-range force. Figure 1.11 depicts such a trajectory: the particles can approach each other but they never really meet.

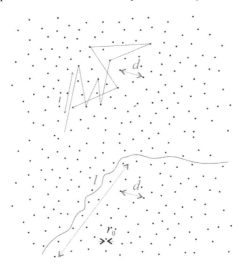

Figure 1.11. *Electron (or ion) trajectory shape in a plasma (bottom). The characteristic distances are represented: l is the mean free path, d is the average distance between particles and r_0 is the Landau length. (Top) For comparison, the situation in a neutral gas; the mean free path is the average distance traveled between two shocks of billiard ball type. For a color version of this figure, see www.iste.co.uk/belmont/plasma.zip*

Among these collisions, both close and distant collisions occur *a priori*. In fact, the weight of the latter is much greater and the close (binary) collisions are negligible in a weakly correlated plasma (which is the general framework of this book). To characterize *the close interaction*, we can calculate the Landau length. This distance is that at which two particles must approach to have interaction energy equal to their average kinetic energy, $k_B T$, where k_B is the Boltzmann constant and T

is the temperature. If we consider an electron, the Landau length is $r_0 = \dfrac{e^2}{4\pi\varepsilon_0 k_B T}$; it is in some way the particle size if considering the electrostatic interaction. This allows us to define an effective section $\sigma = \pi r_0^2$ and the mean free path for the close interaction:

$$l = \frac{1}{n\sigma} = \frac{16\pi\varepsilon_0^2 (k_B T)^2}{ne^4}$$

where n is the particles' density in the system. The higher the density, the smaller the mean free path and the more important the collisions. On the other hand, high temperatures correspond to large mean free paths, and therefore to few collisions.

A more in-depth estimate[1], which also takes into account *distant interactions*, is based on Rutherford scattering and calculates the charged particle from all other plasma particles. It shows that the collision cross-section increases and is about $\sigma \approx 4 \ln\Lambda \, \pi r_0^2$, where $\ln\Lambda$, called Coulomb logarithm, is given by $\Lambda \propto n\lambda_D^3$. This parameter Λ takes into account the distant interactions by assuming a screening phenomenon at the Debye length (with $\lambda_D \gg r_0$). It is very large in weakly correlated plasmas: $\ln\Lambda$ varies from 5 to about 30. With this assumption, the calculation of the mean free path is as follows:

$$l = \frac{4\pi\varepsilon_0^2 (k_B T)^2}{ne^4 \ln\Lambda}$$

Comparing this result with the mean free path calculation based only on close interactions in the previous paragraph, we note that the long-range effect of the Coulomb force is to reduce the mean free path by a factor of $4 \ln\Lambda$ (factor of about 20–120). More detailed considerations on the notion of collision will be given in Chapter 3.

The lines drawn in fine lines in Figure 1.5 give the typical values of mean free path according to plasma characteristics. It is found that for most astrophysical plasmas, l is of the order of magnitude of considered regions' size. Collisions are therefore very often completely negligible, as soon as the phenomena studied are of a smaller scale than this global scale.

In plasma of density n, the average distance d between particles is $n^{-1/3}$. The two distances d and r_0 can be compared as a function of density and temperature. The blue line drawn in Figure 1.5 is obtained for $d = r_0$. It can be seen that almost all the plasmas are situated on the left-hand side of the line ($r_0 < d$) and are so diluted that

---

1 For detailed calculation, see, for example, (Delcroix and Bers 1994, Chapter 3).

the probability of a close binary encounter is totally negligible. In the opposite side of the spectrum, we have strongly correlated plasmas: the potential energy of interaction between particles is then greater than the average kinetic energy. This case will not be dealt with in this book.

We therefore mainly assume the following inequality: $r_0 \ll d \ll \lambda_{De}$. These are plasmas for which we have a large number of particles in a sphere of radius equal to the Debye length ($n\lambda_{De}^3 \gg 1$ or $d \ll \lambda_{De}$). This allows us to introduce a parameter called the plasma parameter:

$$g = \frac{1}{n\lambda_{De}^3} = \frac{4\pi r_0}{\lambda_{De}} = \left(\frac{d}{\lambda_{De}}\right)^3 \propto \left(\frac{r_0}{d}\right)^{3/2}$$

This parameter is small ($g \ll 1$). The line $\lambda_{De} = d$ is shown in Figure 1.5 (yellow line). We see that most of the plasmas mentioned are on the left-hand side of this line and thus fall into this category of weakly correlated plasmas. For these plasmas, the collective motion of charged particles dominates, and we can neglect the effect of collisions between particles, that is, the effect of thermal noise on the trajectories. If we have the opposite situation, that is, $g \gg 1$, then we will get a situation where the plasmas are strongly correlated: indeed, we can easily show that $g \gg 1$ implies $r_0 \gg d$.

1.3.3. *Role of quantum effects*

Plasma electrons are fermions and we know that if they find themselves too close to each other, the quantum effects will intervene. These effects can be characterized by the Fermi energy:

$$\epsilon_F = \frac{h^2\left(3\pi^2 n\right)^{2/3}}{8\pi^2 m_e}$$

The comparison between thermal energy and Fermi energy is equivalent to the comparison between the distance d between particles and the de Broglie wavelength; it leads to the drawing of the red line in Figure 1.5. It is recalled that the de Broglie wavelength for a set of electrons at temperature T is defined by $\lambda_{dB} = h/\sqrt{k_B T m_e}$. It can be seen that for most plasmas, on the left-hand side of the line, the thermal energy is several orders of magnitude higher than the Fermi energy. The point that represents a typical metal is well below this line, which corresponds to the well-known fact that quantum effects are fundamental for understanding metallic behavior. Plasmas in the interior of some stars may be sensitive to quantum effects, and they are thus called degenerates. They will not be studied in this book.

The conclusion that can be drawn from this rapid study is that the effects governing plasma physics are, in general, fundamentally different from those governing gases. The binary interactions are often negligible and only the collective and quasi-collective effects then intervene in the dynamics: these will be studied in the continuation of this work.

1.3.4. *Role of the magnetic field*

Since plasma consists of charged particles, it is sensitive to the magnetic field. We will see in Chapter 2 the role played by this field on the trajectories of particles. This effect is all the more important as the field is strong. In this case, we are talking about highly magnetized plasma. The influence of the magnetic field is quantified by introducing the parameter β, ratio of the kinetic pressure of the particles to the magnetic pressure:

$$\beta = \frac{\sum nkT}{B^2/2\mu_0}$$

1.4. Coupled field/particle system: general case

In plasma, the variations of the electromagnetic field and the movements of the charged particles are coupled. Generally:

– if we know the **E** and **B** fields, then we know the electromagnetic forces exerted on the particles and how these forces influence the trajectories;

– if we know the position and velocity of all the particles of the plasma, electrons and ions, then we can deduce the charge density ρ and the current **j**, which are the source terms, in the Maxwell equations, of the **E** and **B** fields.

This coupled and self-consistent system is the basis of all plasma physics. This science lies at the intersection of electromagnetism (Maxwell) and statistical mechanics (how to go from information on individual trajectories to the macroscopic information that are ρ and **j**). This is depicted in Figure 1.12.

It is important to note that fields **E** and **B** are the so-called "collective" fields, that is, averaged in some manner (which will be specified in Chapter 3). To these collective fields is actually added, in the instantaneous field measured at a point, a "thermal noise" of fluctuations very rapidly varying in time and space. This noise is due to the particulate character of the plasma: it is included here in the form of what is noted as "collision forces" in the particle motion equation.

The large loop of this figure does not exist in the neutral gases: the two squares in the top right and bottom left are then not connected by any arrow, and the dynamics of the particles and electromagnetic fields are decoupled.

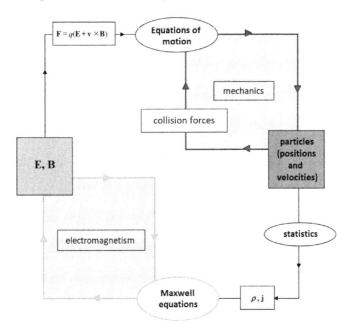

Figure 1.12. *Coupling between particle movements and variations of the electromagnetic field in a plasma. For a color version of this figure, see www.iste.co.uk/belmont/plasma.zip*

1.5. Special case: plasma oscillation

1.5.1. *Position of the problem*

We consider a homogeneous plasma, composed of motionless ions and electrons (so each species has a zero average velocity and a zero temperature) with the same numerical densities $n_e = n_i = n_0$ (and therefore no charge density) and without an electromagnetic field. Without external disturbance, this system obviously remains stationary.

At time $t = 0$, a slice of this system is disturbed by compressing the electrons slightly in the x direction (see Figure 1.13). We thus start from an initial situation where $n_e = n_0 + n_1(x)$ and for which the local charge density is no longer zero: $\rho = e(n_i - n_e) = - en_1 \neq 0$.

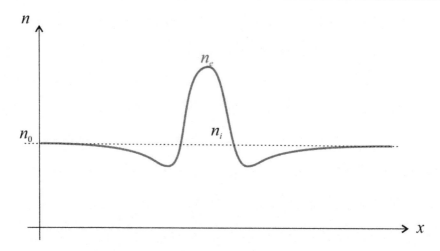

Figure 1.13. *Initial perturbation of the electron density*

How will the medium respond to this perturbation of the electron density? Can the disturbance propagate as a density variation in a collisional neutral gas (sound wave) would do? If so, with what characteristic speed? Or will it be deformed, damped? Will it oscillate on the spot? These are the questions we will attempt to answer. Let us write the evolution questions for this purpose and solve them.

1.5.2. *Field created by electric charges*

The charge density can be set equal to en_1 if it is assumed that the numerical density of the ions remains constant. This hypothesis can be justified *a posteriori* because the evolution that we will find is made on a timescale fixed by the electrons, which is such that the ions have hardly the time to move (the protons are approximately 2,000 times more heavy as electrons, and the other ions are even heavier).

The electric field created by this non-zero charge density is directed along x and is calculated by the Maxwell–Gauss equation:

$$\partial_x \left(E_x \right) = -en_1 / \varepsilon_0 \qquad\qquad [1.1]$$

(noting ∂_x the partial derivation operator with respect to x: $\partial_x(A)$ or $\partial_x A = \partial A / \partial x$).

NOTE.– There is no associated magnetic field!

1.5.3. *Displacement of charged particles due to the field*

The electrons move under the effect of the electric field. Newton's second law applied to each electron is written as follows:

$$m_e d_t v_{ex} = -eE_x$$

The displacement of electrons is obviously from the densest regions to the less dense regions. To calculate the effect of these individual movements on the evolution of the density (and thus to complete the system), a statistical calculation is generally required. Here, this work is reduced to its simplest expression: since the electrons were supposed to be cold, they all start with the same initial velocity $v_{ex} = 0$, and all those originating from the same point x follow the same trajectory and see the same field at the same time. The individual velocities of the electrons in a given x are therefore all equal to each other and equal to the macroscopic velocity at this point:

$$v_{ex} = <v_{ex}> = u_{ex}$$

We can thus derive an equation for the fluid velocity variable:

$$m_e[\partial_t u_{ex} + u_{ex}\partial_x u_{ex}] = -eE_x$$

The variation of density resulting from these movements is deduced from the conservation equation of the number of particles or "continuity equation" (which will be demonstrated in Chapter 4):

$$\partial_t n_e + \partial_x(n_e u_{ex}) = 0$$

If we consider small perturbations, that is, $n_1 << n_0$, then we can linearize these two equations and write:

$$m_e \partial_t u_{ex1} = -eE_{x1}$$

$$\partial_t n_1 + n_0 \partial_x(u_{ex1}) = 0$$

where we used the fact that n_0 is constant and neglected the term which is a product of two small perturbation terms because it is very small compared to these ones. The result is therefore, by combining the two equations:

$$\partial_t^2(n_1) = \frac{n_0 e}{m_e}\partial_x(E_{x1}) \qquad [1.2]$$

1.5.4. *Coupled system resolution*

The set of the two equations [1.1] and [1.2] forms a system in n_1 and E_x. If we solve, for example, n_1, we obtain:

$$(\partial_t^2 + \omega_{pe}^2)(n_1) = 0$$

with $\omega_{pe}^2 = n_0 \dfrac{e^2}{m_e \varepsilon_0}$.

We recognize the equation of an oscillator whose general solution is:

$$n_1(x,t) = n_{1a}(x)\cos(\omega_{pe}t) + n_{1b}(x)\,\sin(\omega_{pe}t)$$

The characteristic frequency ω_{pe} is called the electron[2] plasma frequency. In each slice in x, the electron density perturbation oscillates without damping (within the framework of the approximations made) with this unique frequency ω_{pe}.

The oscillation phase of each slice is independent of its neighbor and is fixed by the initial conditions. If we fix for n_1, for example, an initial spatial variation of sinusoidal shape and characterized by a wave number k, we will generally obtain the superposition of two sinusoids propagating one to the right, and the other to the left. The corresponding amplitudes of these two sinusoids are fixed by another initial condition, for example, the derivative $\partial_t(n_1)$ or the value of the electric field. If the two amplitudes are equal, then no propagation is observed, but a stationary wave; this is what we get from an immobile initial condition (all slices then oscillate in phase).

We have just seen that the oscillation frequency $\omega = \omega_{pe}$ of each slice is independent of the shape chosen for the spatial variation; for a sinusoidal spatial variation, it is thus *a fortiori* independent of the wave number. By using the wave vocabulary, which is very useful for characterizing the properties of a medium, this means that the *dispersion relation* $\omega(k)$ here takes on the remarkably simple form illustrated in Figure 1.14.

---

2 NOTE.– although ω_{pe} is a pulsation (rad / s) and we denote the frequency by fpe = $\omega_{pe}/2\pi$, it is customary to speak in both cases of frequency, even if strictly speaking we should make a difference between frequency and pulsation.

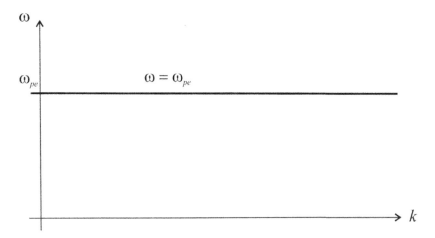

Figure 1.14. *Dispersion relation ω(k) of the plasma oscillation*

The wave thus found is sometimes called "Langmuir wave". The zero slope of the dispersion relation $\partial\omega/\partial k = 0$ shows the absence of energy propagation (zero group velocity). On the contrary, we see that the phase velocity $\omega/k = \omega_{pe}/k$ depends on k and can take all the values between zero and infinity. Note the difference between the above result, obtained when calculating the evolution of a density variation in a neutral and hot gas (sound wave). In this case, we obtain a linear dispersion relation $\omega = kc_s$. The frequency of oscillation then depends on the spatial form, and, in this case, group velocity = phase velocity = constant. (Each slice of fluid has an oscillation that depends on the oscillation of the neighboring slices, which pass the information to it with a slight delay.)

It has just been shown that there can exist, in a plasma, "electrostatic waves", that is, waves for which the electric field has only a longitudinal component (i.e. along the direction x of **k**, which is the direction of all the gradients and therefore the direction of propagation). These phenomena are called "electrostatic" because the longitudinal component of the field is due to a charge density that does not involve a magnetic field, much different from the usual electromagnetic waves in the vacuum, where the electric polarization is purely transverse.

1.6. Plasma frequency

1.6.1. $\tau = \omega_{pe}^{-1}$: characteristic electron response time

This is the first characteristic time of a plasma that we encounter (it is not the only one; we will see others that are associated with other phenomena). This is the response time of electrons to the deviations to neutrality: in a cold plasma (i.e. where we neglect the effects of temperature), we have just seen that an excess of charge cannot exist in a stationary manner over a time greater than or equal to ω_{pe}^{-1}: under the effect of the electrostatic field that acts as a restoring force, the electrons rush to fill the holes in -V (opposite of the electrostatic potential) in a time of about τ (then, carried by the momentum, they exceed the position of balance and everything starts again in the other direction. It is the same principle as a spring that oscillates).

Moreover, the plasma frequency also appears as one of the privileged frequencies where the exchanges of energy between the electromagnetic field and the thermal energy of the particles can occur. It plays a particular role for the amplification or the absorption of waves (see instability of beam, phenomenon of fading of the radio waves due to the ionosphere, heating of the plasma, inertial fusion, etc.).

NOTE.– In cold and non-magnetized plasma, only the characteristic time $\tau = \omega_{pe}^{-1}$ appears. More generally, it is clear that there are many other phenomena than plasma oscillations in all situations of less idealized plasma physics than that which was just studied (with temperature effects, in the presence of currents and magnetic field, with movement of ions, nonlinear situations, 3D geometry, etc.), which means that in plasmas, there is a real "zoo" of possible waves! However, it remains true that, in all cases, this frequency organizes the behaviors: for all the phenomena occurring at lower frequencies ($\omega << \omega_{pe}$), the electrons can be considered as infinitely fast, ensuring quasi-neutrality of the plasma (except at spatial scales smaller than λ_D if the plasma is hot), so partly canceling the electrostatic field. On the contrary, beyond this frequency ($\omega \geq \omega_{pe}$), the inertia of the electrons must be taken into account and all the phenomena involve deviations from quasi-neutrality (and the additional electrostatic field corresponding to it).

1.6.2. Some characteristic plasma frequencies

We have seen that, by definition, $\omega_{pe}^2 = n_0 \dfrac{e^2}{m_e \varepsilon_0}$. Numerically, the corresponding frequency $f_{pe} = \omega_{pe}/2\pi$ is (f_{pe} in Hz, n_0 in m^{-3}):

$$f = 9\sqrt{n_0}$$

If n_0 is expressed in cm^{-3}, then the same formula gives the plasma frequency in kHz.

Here are some examples of orders of magnitude (the wavelength λ_o of the waves in vacuum for electromagnetic waves with $\omega = \omega_{pe}$ is given for reference, *but beware: it is not the wavelength of the wave in the plasma*):

	Electron density (m − 3)	Plasma frequency (λ_0 = c/fpe)
Lobes of the magnetosphere	10^4	900 Hz ($\lambda_0 \approx$ 300 km)
Solar wind	5×10^6	20 kHz ($\lambda_0 \approx$ 15 km)
Max ionosphere (250 km)	5×10^{11}	6 MHz ($\lambda_0 \approx$ 50 m)
Solar corona	10^{14}	90 MHz ($\lambda_0 \approx$ 3 m)
Atmosphere of pulsar	10^{18}	9 GHz ($\lambda_0 \approx$ 3 cm)
Discharge tube (neon)	10^{18}	*Idem* (but partial ionization)
Fusion plasma (tokamak)	10^{21}	3×10^{11} Hz (IR)
Laser-created plasma	10^{27}	3×10^{14} Hz (visible)
Electrons in a metal	10^{29}	3×10^{15} Hz (UV)
Interior of the Sun	10^{32}	9×10^{16} Hz (UV)

Table 1.1. *Plasma frequencies in some characteristic environments. The wavelengths indicated in parentheses correspond to the electromagnetic waves in the vacuum that would have the same frequency*

1.7. Effects of temperature

When one or more of the simplifying hypotheses used in the above processing are removed, it is found, on the one hand, that the properties of the oscillation of plasma are modified compared to the previous results, and, on the other hand, that there is a large number of other eigenmodes of the plasma (i.e. natural oscillation frequencies associated with the new phenomena taken into account). In this introductory chapter, we will only mention the thermal motion of the particles. The possibility of exciting other waves in the plasma and the detailed calculation of the change in plasma oscillation will be presented in Chapters 4 and 6.

1.7.1. *The velocity distribution function*

If the electron gas is not "cold", it means that the initial velocities v_{eo} of the different particles are not zero. Considering that the number of electrons is large enough for us to be able to use a statistical treatment (this is a hypothesis always verified in practice), the electron gas can be characterized by its "distribution function". The probability distribution of the velocities of the set of electrons is thus sought; we can, of course, define it for any moment t (not necessarily the initial moment $t = 0$) and any given position x. The graphical representation of this velocity histogram gives the appearance of the distribution function; an example is shown in Figure 1.15.

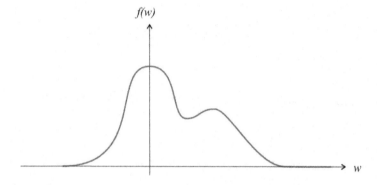

Figure 1.15. *Example of a velocity distribution function*

The Gaussian distribution function plays a special role because it is the equilibrium function for neutral collisional gases. It is more often called "Maxwellian" in this context, with reference to the Maxwell–Boltzmann distribution, well-known in statistical physics. If the collisions are negligible, then we will see that the distribution function of the plasma can significantly deviate from the Maxwellian distribution.

Once the particle distribution function has been identified, it is useful to know how to deduce the usual macroscopic quantities (density, fluid velocity, pressure, etc.). These quantities constitute, in the sense of probability distributions, the "moments" of the distribution function. They make it possible to know, quantitatively but without knowing all the details of the function, whether this distribution is centered around a zero or non-zero average velocity, whether it is wide or narrow and so on. The following are the definitions of the first moments; their expressions are simplified here thanks to the one-dimensional context in which we work in this introductory chapter. Full vectorial (and tensorial) definitions will be given in Chapter 4.

Numerical density: $n = \int\limits_{-\infty}^{\infty} f(v)\,dv$

Fluid velocity: $u = <v> = \dfrac{1}{n}\int\limits_{-\infty}^{\infty} v\,f(v)\,dv$

Mean quadratic speed: $V_{th}^2 = <(v-u)^2> = \dfrac{1}{n}\int\limits_{-\infty}^{\infty} (v-u)^2\,f(v)\,dv$

Third moment (heat flow): $\dfrac{q}{nm} = <(v-u)^3> = \dfrac{1}{n}\int\limits_{-\infty}^{\infty} (v-u)^3\,f(v)\,dv$

The definitions of the kinetic pressure and the temperature are deduced from the previous thermal velocity: $p = nmV_{th}^2 = nk_BT$.

In the particular case of a Maxwellian distribution, the mathematical expression of f is (in one dimension):

$$f(v) = \dfrac{n}{\sqrt{2\pi}V_{th}} e^{-\frac{(v-u)^2}{2V_{th}^2}}$$

and it is quite clear that the first three moments, n, u and V_{th}, are sufficient to completely determine the form. If need be, we can calculate all the other moments according to the second (V_{th}): the even moments are given by $<v^n> = (n-1)!!V_{th}^n$ and the odd moments are null[3]. This explains why all the fluid theories (hydrodynamics, gas theory, thermodynamics, etc.) used in collisional media are based on just three macroscopic quantities: density, velocity and pressure. This is not true in general for plasma without collisions where the restriction of a distribution function at only a few first moments must be done with caution: in general, the distribution function is equivalent to the data of the infinite series of its moments, and any break in this series is always an approximation.

---

3 The double factorial is defined by $p!! = p \times (p-2) \ldots \times 3 \times 1$ for odd p.

1.7.2. The different theoretical treatments

To treat a plasma physics problem such as that of plasma oscillation, two methods are used:

– *kinetic treatment*: the complete treatment. We calculate the evolution of the distribution function $f(v)$ as a function of x and t using a kinetic equation such as the Vlasov equation (see Chapter 3), and we deduce (thanks to the moments we have just introduced) the macroscopic quantities of interest, n, u, p (or T), E, B, etc. This treatment will be discussed in Chapter 6 for the case of plasma oscillation. It should be noted that in general, this complete treatment is difficult;

– *fluid treatment*: we simplify the problem by reducing it (through integrations of the kinetic equation) to a system of equations that shows only the first few moments (usually n, u and p or T) and solving it. This is quite easily achievable in the collisional case. However, for a collisionless environment, it cannot be said that such a limited system exists in all generality and that the first moments are always connected to each other independently of the higher-order moments. This can only be done when justified approximations allow. This will be detailed in Chapter 4, with the discussion of the notion of "closure" of the fluid system.

Given the complexity of the equation systems to be solved, it is often necessary to use a numerical approach to kinetic or fluid treatment. Kinetic treatment remains rather "expensive" in terms of numerical computation and in practice, it is currently only used for somewhat idealized problems, especially with simplified geometries and boundary conditions.

1.8. Some examples of application of plasma physics

1.8.1. Nuclear fusion and plasma physics

Fusion results from the encounter of two light nuclei (positively charged). If we can get them close enough, then they will merge into a heavier nucleus with a very strong release of energy. The Sun is a natural thermonuclear reactor, whose confinement is ensured by the force of gravity and the pressure of the plasma mass surrounding the star's core (Figure 1.16). The fusion of hydrogen (proton–proton reaction) is very slow, and the star remains stable. The temperature is of the order of 15×10^6 K (1.5 keV). On Earth, the simplest reaction that can be produced in the laboratory is the fusion of a deuterium nucleus and a tritium nucleus that produces a nucleus of helium and an energetic neutron according to the reaction $D + T \rightarrow \alpha(3.3$ MeV$) + n$ (14.1 MeV). For this, we need temperatures of the order of 100×10^6 K (10 keV) and the material expected to produce fusion is in the plasma state.

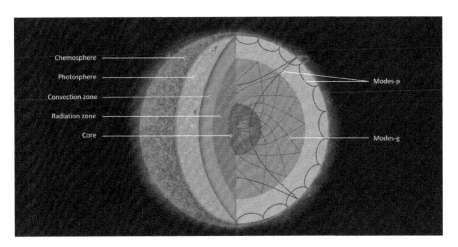

Figure 1.16. *Fusion reactions in the Sun's core prodjuce energy. SOHO observations have made it possible to calculate the dimensions of the core and to calculate its rotation speed, thanks to seismic waves (source: translated SOHO image, ESA/NASA). For a color version of this figure, see www.iste.co.uk/belmont/plasma.zip*

There is a basic requirement for performing the fusion reaction; this is called Lawson's criterion: for plasma of density n, confined for a time τ, the product n tau must be greater than a critical threshold for the fusion to occur. This critical threshold is the lowest for a temperature of about 100 million Kelvin, and it is of the order of 10^{20} sm^{-3}. To keep the material in a plasma state in the laboratory, it must be confined. Two lines of research are currently trying to achieve this confinement in parallel. In the case of magnetic confinement fusion, confinement is achieved through intense magnetic fields produced in machines known as tokamaks (see Figure 1.17). The plasma is tenuous and we try to make the confinement last as long as possible. In these machines, the magnetic field has two components: a circular horizontal, called toroidal, and the other, in the section plane in the form of a D, called poloidal. This complex structure of the field is optimal to avoid plasma drifts (see Chapter 2).

In the case of inertial confinement fusion, plasma is created by heating a ball of dense material with a laser, like, for example, the Megajoule laser (Figure 1.18). We are thus in a reverse situation with respect to Lawson's criterion: the confinement does not last long (the experiment is close to an explosion) and we try to obtain the densest plasma possible, starting from the solid state. Other similar experiments use particle beams and pressure confinement.

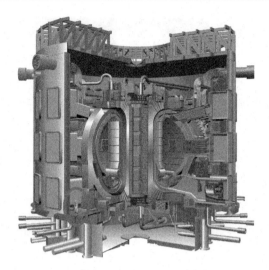

Figure 1.17. *Schematic diagram of the ITER tokamak (magnetic confinement). Inside the red ellipse is the silhouette of a man which demonstrates the scale of the machine. For a color version of this figure, see www.iste.co.uk/belmont/plasma.zip*

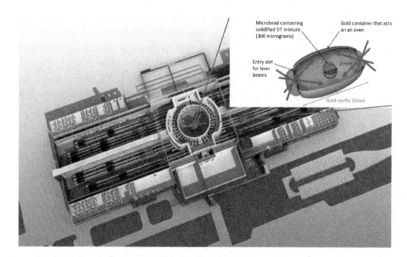

Figure 1.18. *Megajoule laser, CEA (inertial confinement). Schematic diagram of a building that contains 22 laser lines and measures 300 m long. The sidebar shows the 10 mm gold cavity, where the fusion reaction takes place. For a color version of this figure, see www.iste.co.uk/belmont/plasma.zip*

1.8.2. *Cold plasmas and their applications*

The so-called "cold" plasmas, in the sense of the discharge plasmas, cover a wide range of pressures, ranging from 0.1 Pa to a few atmospheres. It can even be created by discharge in liquids.

Considering that the cold plasma is not completely ionized, it is a mixture of molecules, atoms, radicals and reactive ions. It can therefore be the site of chemical reactions, and a number of applications are related to this chemical reactivity, such as de-pollution (Figure 1.19), surface treatments (used to make deposits or etching) and the first applications appearing in the biomedical field. Most cold plasmas are obtained by electric discharge. The best known are found in low-consumption lamps, but they are plasmas that are also found in the phenomenon of lightning. For these lamps and for laboratory or industrial applications, discharges are alternating, at frequencies ranging from a few kilohertz to tens of Megahertz (referred to as radio frequency plasma, see Figure 1.20). There is another application of cold plasmas that is not made to work in the laboratory, which is the so-called "electric" thrusters that propel satellites by ejecting ions and not neutrals as in a conventional chemical thruster (see Figure 1.21).

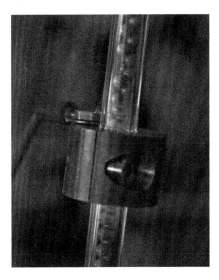

Figure 1.19. *Plasma prototype for air pollution control: the plasma is generated by a dielectric barrier discharge in the presence of a catalyst (beads) (source: C. Barakat, LPP). For a color version of this figure, see www.iste.co.uk/belmont/plasma.zip*

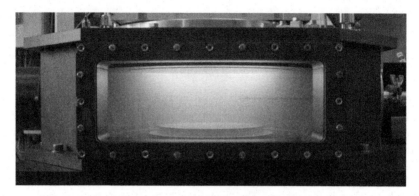

Figure 1.20. *LPP radio frequency reactor used for etching silicon with chlorine plasma (source: J.P. Booth). For a color version of this figure, see www.iste.co.uk/belmont/plasma.zip*

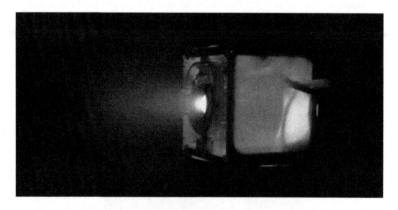

Figure 1.21. *Plasma thruster: on the left-hand side is the plasma ejected by the thruster (source: A. Aanesland, ThrustMe, www.thrustme.fr). For a color version of this figure, see www.iste.co.uk/belmont/plasma.zip*

Is the ionosphere of the Earth cold plasma or hot plasma in the previous sense? It is not usually referred to in these terms. In reality, the behavior of the plasma evolves according to the altitude. In regions E and F (see Figure 1.4), which correspond to the electron density maximum, the plasma is cold with an electron temperature much higher than that of ions and neutrals. When we go up, the density of the neutrals sharply decreases (until they eventually disappear), while the plasma is also rarefied and the temperature of the ions increases to become the same as that of the electrons. The plasma thus becomes hot (like that of the magnetosphere, which is above), in the sense that the ions have temperatures of the same order as electrons, even if this temperature is not very high (of the order of 3,000 K at 1,000 km).

1.8.3. *Production of energy particles by plasma accelerator*

If we focus a laser on a small plasma (micron to millimeter scale), then it is possible to create beams of highly energetic particles. During the interaction of the laser with the plasma, extreme electric fields are produced, which can reach values of the order of the TV/m, more than 10,000 times more intense than the electric fields produced in the radio frequency structures of the "traditional" accelerators. Different schemes can accelerate electrons and/or ions. Figure 1.22 illustrates the acceleration called TNSA (target normal sheath acceleration): the electrons of the plasma heated by the laser form a front of negative charges, and the separation of the charges generates an intense electric field, which accelerates the ions. The thus created ions can be used for different applications, such as medical, nuclear and diagnostic applications, with the advantage of having been created in a very small system.

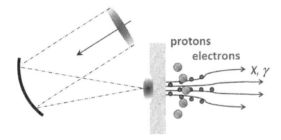

Figure 1.22. *Ion acceleration by target (normal sheath acceleration). The laser beam is focused on a target. The energy of the laser is transferred to electrons and radiation through a very intense current in the plasma. For a color version of this figure, see www.iste.co.uk/belmont/plasma.zip*

1.8.4. *Astrophysical plasmas*

As we have seen, the vast majority of the observable matter in the universe is in the form of plasma: an environment as cold, dense and protected from ionizing radiation as the Earth's surface is an exception. The term "space plasmas" is generally used to refer to all astrophysical plasmas sufficiently close to the Earth to be accessible by space probes capable of performing *in situ* measurements. In order of the distance to Earth, these include the ionosphere, studied by rocket probes from the 1960s, and the magnetosphere, the plasma environment contained in the zone of influence of the Earth's magnetic field (up to about 10 terrestrial radii on the side of the Sun). Beyond the magnetosphere is the interplanetary medium, bathed by plasma radially flowing from the Sun: the solar wind. This wind is created by the evaporation of the solar atmosphere (or corona), too hot to be confined by solar gravity. It blows up to distances of about 100 AU, well after the orbit of the last planet (Neptune is at 30 AU). This sphere of influence of the plasma (and hence of the magnetic field) of solar origin is called the "heliosphere". Beyond extends the interstellar medium, composed of a very heterogeneous mixture of gases in atomic and molecular forms, and plasma. The stars belong to this interstellar medium. Their plasma environment can be comparable to that of the Sun (stars of the main series), but can also be very different: there are, for example, highly magnetized environments around compact objects such as neutron stars, responsible for the radio-astronomical phenomena of pulsars (fields on the surface reaching 10^{11} T!). The heart of these stars provides an example of degenerate plasma, where quantum effects play a crucial role.

To conclude this chapter, Figure 1.23 summarizes the wide variety of plasmas and applications in the study of plasma physics.

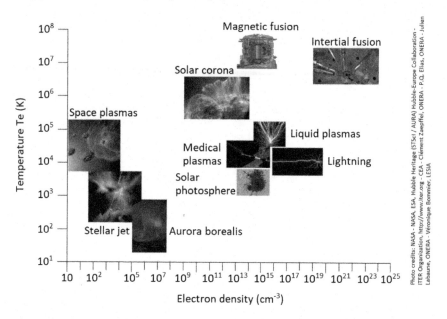

Figure 1.23. *Density/temperature diagram for different types of plasmas (source: adapted from a figure from labexPlas @ Par). For a color version of this figure, see www.iste.co.uk/belmont/plasma.zip*

2

Individual Trajectories in an Electromagnetic Field

We have seen that the knowledge of the individual trajectories of particles is a fundamental link in the study of the self-consistent system called plasma. Before any statistical study, let us begin by specifying what these trajectories are in the presence of given magnetic and electric fields. This study constitutes the first step to describe the magnetized plasmas. Then, we can try to close the system and to calculate how the knowledge of all the particle trajectories can determine the fields themselves. Here, we will consider trajectories without collision.

2.1. Trajectory of a particle in a uniform and stationary magnetic field

Let us begin by briefly recalling known results on the trajectory of a charged particle in the simplest case: a constant electromagnetic field without any force (gravity, collisions, etc.). Remember that here we note $\mathbf{v}$ as the velocity of the individual particles.

2.1.1. *Without electric field (E = 0)*

In this case, a charge particle q is only subject to the magnetic component of the Lorentz force, $\mathbf{F} = q\mathbf{v} \times \mathbf{B}$. Its trajectory is obtained without difficulty by integrating the equations of motion and consists of a helix wound around a **B**-field line:

– parallel to the magnetic field: uniform rectilinear motion $v_\parallel$ = cste;

– perpendicular to the magnetic field: uniform circular motion: $v_\perp$ = cste;

where the parallel and perpendicular components (with respect to the magnetic field) of the velocity are defined as $\mathbf{v}_\parallel = \mathbf{v} \cdot \mathbf{b}\,\mathbf{b}$ and $\mathbf{v}_\perp = \mathbf{v} - \mathbf{v}_\parallel$, respectively, with $\mathbf{b}$ being the unitary vector collinear with the magnetic field vector.

The gyration motion in the plane perpendicular to $\mathbf{B}$ is called the cyclotron motion (Figure 2.1), which is characterized by:

– its rotation frequency (*gyro-frequency*): $\omega_c = 2\pi f_c = qB/m$ (independent of the particle's energy);

– its radius (*Larmor radius*): $\rho_L = v_\perp / \omega_c$ (proportional to $v_\perp$ and therefore dependent on energy).

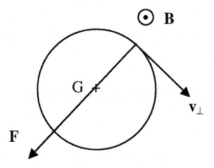

Figure 2.1. *Positive-ion motion in the plane perpendicular to B*

Note that a *positive* ion rotates in the "left" direction around the magnetic field (clockwise direction in the plane of the figure oriented by $\mathbf{B}$).

We call the angle α pitch angle, which is the angle that makes the velocity with $\mathbf{B}$. This angle, which is constant in the present case, is one of the important characteristics of the trajectory since it determines "the pitch" of the helix. It is connected to the parallel and perpendicular components of the velocity by:

$$tan\, \alpha = \frac{v_\perp}{v_\parallel}$$

The pitch of the helix, which means the distance traveled along the magnetic field during a complete Larmor rotation, is therefore:

$$h = v_\parallel (2\pi/\omega_c) = 2\pi \rho_L \cot \alpha$$

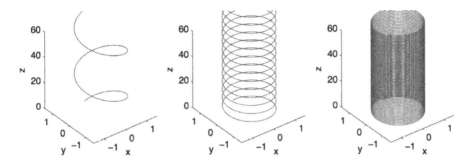

Figure 2.2. *Motion of a positively charged particle in a magnetic field oriented along the Oz-axis, for different values of pitch angle (from right to left, α = 10°, α = 45° and α = 80°). Distances are normalized by the Larmor radius of the particle*

If we consider the average motion of a particle on a long timescale with respect to the gyro-period, then this movement merges with that of the rotation center G, which is called the guiding center. In the considered simple case, the guiding center's motion is therefore characterized by:

– parallel to the magnetic field: uniform rectilinear motion $v_{G\parallel} = v_{\parallel}$ = cste;

– perpendicular to the magnetic field: no motion $v_{G\perp} = 0$.

This last result, although obvious, is important for the following reason: in the absence of an external force field, the mean perpendicular motion of a charged particle is zero; the particle remains "attached" to the same field line around which it rotates. If we propel the particle to the right with a higher initial velocity, then it will not acquire a mean velocity to the right but it will only have a larger gyration radius (and the rotational angular velocity will not change).

2.1.2. *With a stationary and uniform electric field (E ≠ 0)*

2.1.2.1. *Motion parallel to the magnetic field*

The Laplace force $\mathbf{F} = q\mathbf{v} \times \mathbf{B}$ has no component in the B-field direction. The particle can only be subjected to the parallel component of the electric field $E_{\parallel}$, if it exists. This would then provide a constant acceleration inversely proportional to its mass. This result, *in plasma*, is more on the academic side; practically, such a uniform and stationary $E_{\parallel}$ component cannot exist, unless it is equal to and opposite to another force (e.g. gravity): according to Chapter 1, regardless of whether particles of plasma are subjected to parallel forces of non-zero resultant, the electrons' motion, faster than that of the ions, would lead to a separation of charges, which would be an immediate source of a new parallel field coming to oppose the external force. Depending on the nature of the boundary conditions, this would either return to the stationary value $F_{//} = 0$ or produce an oscillation at the plasma frequency. In the case $F_{//} = 0$, the parallel motion is, as in the previous case, at a constant velocity.

2.1.2.2. Perpendicular motion

To compute this trajectory, we could of course write the fundamental equation of dynamics $\mathbf{F} = m\mathbf{a}$ in the considered reference frame R and thus interpret all local curvature changes. However, we can find a much faster way to proceed if we remember first that the result is already known in the absence of an electric field and second that the value of the electric field depends on the considered reference frame.

If we calculate the trajectory in a reference frame R' moving at a velocity $\mathbf{V}_{R'}$ with respect to the preceding one, then the reference change, for non-relativistic velocity, does not change the magnetic field, but it changes the electric field by:

$$\mathbf{E}' = \mathbf{E} + \mathbf{V}_{R'} \times \mathbf{B}$$

Among all possible reference frames, it is natural to choose a reference frame where $\mathbf{E}'_{\perp} = 0$ since it is in this reference frame that the perpendicular trajectory is known and is particularly simple. For this purpose, the reference frame $R' = R_m$, called the "magnetic" reference frame, is defined as being that which moves with respect to R at the perpendicular velocity $\mathbf{V}_m$ such that $E_{\perp m} = 0$, that is:

$$E_{\perp m} = 0 \iff \boxed{\mathbf{V}_m = \frac{\mathbf{E}}{B} \times \mathbf{b}}$$

If R_m has a parallel velocity, the value of the field is not changed, so it is simply taken equal to zero.

In the reference frame R_m, the perpendicular trajectory is a uniform circular motion; the particle has no average motion, and its guiding center is motionless. In the initial reference frame R, the trajectory is thus the combination of a uniform circular motion and a translational motion at a constant velocity $\mathbf{V}_m$ of reference frame R_m. It is said that the guiding center, which is therefore no longer stationary in the presence of $\mathbf{E}_{\perp}$, undergoes a "drift" at velocity $\mathbf{V}_m$.

Giving $\mathbf{E}_{\perp}$ or $\mathbf{V}_m$ is equivalent. Depending on whether the value of the electric field $\mathbf{E}_{\perp}$ is greater or smaller, the previously found circular motion somewhat rapidly "drifts" in the direction perpendicular to $\mathbf{E}_{\perp}$. The trajectory is then described by a *trochoid* curve as shown in Figure 2.3.

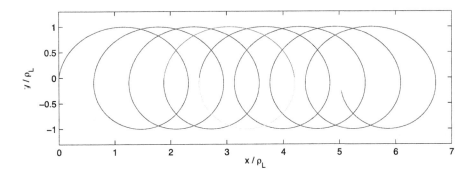

Figure 2.3. *Trajectory projection in the plane perpendicular to B (here along z). To the uniform rotation motion (in the "magnetic" coordinate system) is added a uniform translational motion ("drift" motion). In the figure's frame of reference, the electric field is along y, and is responsible for the variations of curvature[1]: we see that the radius of curvature is greater on the part y > 0 of the trajectory (red) than on the part y < 0 (green). For a color version of this figure, see www.iste.co.uk/belmont/plasma.zip*

This result is reminiscent of similar properties encountered for other rotating objects such as gyroscopes: when a force directed along y is applied, it does not result in an accelerated motion following y, but a uniform motion along x.

SUMMARY.– The velocity of the magnetic reference frame $\mathbf{V}_m = \dfrac{\mathbf{E}}{B} \times \mathbf{b}$ (where the electric field is zero) is an intrinsic quantity of the electromagnetic field. The particles of plasma are not responsible for its existence. Nevertheless, the particles have the property of *materializing* this reference frame since (at least for uniform and stationary fields) it is in this frame that all the particles, regardless of their mass, energy and charge (including the sign of this charge), have circular trajectories in the perpendicular plane and therefore guiding centers, which remain attached to the same field line.

2.1.3. *Constant force*

If the particle is subjected to a constant force other than an electric force (gravity or other), then the results are necessarily similar to those just presented, simply replacing $\mathbf{E}$ with $\mathbf{F}/q$ (although the interpretation in terms of magnetic reference

---

1 In the descending parts of the trajectory ($v_y < 0$), the particles are slowed by the electric field E_y. This gradually decreases their curvature radius (which is proportional to the velocity) to the lower end of the trajectory. The opposite occurs in the "rising" parts. This results in the calculated drift of the center guide towards x > 0.

frame is obviously no longer valid). A particle subjected to a force **F** will thus undergo a drift in the plane perpendicular to the magnetic field, at a constant velocity given by:

$$\mathbf{v}_{dF} = \frac{\mathbf{F}}{qB} \times \mathbf{b}$$

For a force independent of the charge (unlike **F** = q**E**), such as a gravitational force, the corresponding drift velocity depends this time on q, in particular on its sign. The ions and the electrons therefore drift in opposite directions, and hence the existence of currents and eventually of charge separations (and therefore of an electric field in the drift direction).

We have just seen that the particles of magnetized plasma placed in a uniform gravitational field will drift, in the direction perpendicular to both **B** and **g**, with a velocity of mg/qB. This drift being proportional to the mass, we can *a priori* neglect the drift of the electrons compared to that of the ions. Let us calculate the order of magnitude of this drift for different devices. First of all, in a tokamak: the magnetic field has an order of magnitude around a tesla (5.6 T in the case of ITER), and therefore $v_{dg} \simeq 0.1 \mu m/s$, which is of course negligible compared to the velocities involved. In the ionosphere, the drift velocity is a little larger: for B = 10^{-5} T, it is of the order of 0.1 m/s for an ion of mass number A = 10. This remains negligible in most cases.

2.2. Slowly variable fields

The previous result showed that a particle of velocity $(\mathbf{v}_{\parallel}, \mathbf{v}_{\perp})$ in a stationary and uniform magnetic field describes a helix in the coordinate system where **E** = **0**. Let us now look at how this trajectory is modified if the magnetic (or electrical) field is not exactly uniform, or if it is not exactly stationary. This is typically the first term of a perturbation series near a known solution. The development will be valid if the perturbation is "small": it will be seen later that the conditions for this are (written here in a compact and symbolic way):

$$\partial_t \ll \omega_c, \; \partial_{//} \ll \frac{\omega_c}{v_{//}} \; \text{and} \, \nabla_{\perp} \ll \frac{1}{\rho_L}$$

where ρ_L is the Larmor radius. In other words, the results obtained on the drifts will be valid in the case of a slowly variable field with respect to the spatial and temporal scales of the cyclotronic motion.

We will see that these slow variations have two types of consequences that we will treat successively:

– the particle "drifts" in the plane perpendicular to **B** instead of remaining attached to the field line of the magnetic marker;

– energy *exchanges* occurring between the *parallel* motion and the *perpendicular* motion (respecting $v_\parallel^2 + v_\perp^2 = $ cste provided that the parallel electric field is zero).

2.2.1. Drift in a non-uniform B-field, with E = 0

Two types of non-uniformity of the magnetic field are distinguished: the curvature (the direction of **B** changes; the force lines are curved) and the gradients (the modulus of the magnetic field changes, the force lines converge).

2.2.1.1. Curvature drift

If the direction of the magnetic field is not constant, then the parallel motion of the particle induces a centrifugal force due to the curvature of the force lines. This inertial force acts on the perpendicular motion as a constant force as long as the radius of curvature is sufficiently large ($R_c \gg v_\parallel / \omega_c$). If all the other parameters are constant, then this force leads to a drift by the already given law:

$$v_{dc} = \frac{\mathbf{F_c}}{qB} \times b$$

with the expression of the centrifugal force $\mathbf{F_c} = -\dfrac{mv_\parallel^2}{R_c} n$ (where n is the unitary vector perpendicular to the trajectory, oriented towards the inside). Figure 2.4 shows the motion of a positive charge particle in a constant modulus magnetic field, directed in the orthoradial direction (the field has the same direction as if it were created by a current running an axis line Oz, directed "upward"): the field lines are concentric circles centered on the Oz-axis. A particle at a distance R from the Oz-axis will therefore have a drift velocity:

$$\boldsymbol{v_{dc}} = \frac{mv_\parallel^2}{qBR} \boldsymbol{e_z}$$

As shown in Figure 2.4, the trajectory of the particle guiding center is in this case helical. The guiding center revolves around the Oz-axis with a period $2\pi R / v_\parallel$ and has a constant velocity v_{dc} along z. The helix pitch is therefore:

$$h = v_{dc} T = 2\pi \frac{mv_\parallel}{qB} = 2\pi \rho_L \cot \alpha$$

that is, exactly the same pitch as that of the helix of the cyclotron trajectory of the particle around the field line.

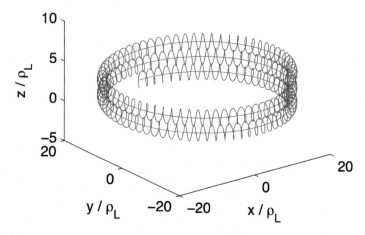

Figure 2.4. *Trajectory of a positively charged particle (in blue) in a "toroidal" field of constant modulus; the lengths are normalized by the Larmor radius of the particle, and the helical trajectory of the guiding center is drawn in red, due to the drift of curvature. Here, the curvature radius of the trajectory is R = 20 ρL. For a color version of this figure, see www.iste.co.uk/belmont/plasma.zip*

Among the applications in which the curvature drift plays an important role, we can mention the confinement of particles in a device like a tokamak, which cannot be effectively carried out by a purely toroidal magnetic field: we would have losses of particles on the "high" and "low" walls due to the drift of curvature (and of gradient).

The solution to this problem involves the addition of the so-called poloidal magnetic field, which "twists" the field lines and compensates the effect of this drift on a complete revolution.

The radiation belts of Earth (and of other magnetized planets) consist of very energetic particles (up to about 400 MeV for protons), which are trapped in the Earth's magnetic field at altitudes ranging from 1,000 km to 60 000 km from the Earth's surface. This magnetic field being oriented from the south to the north, and the centrifugal force towards the outside of the Earth, the positively charged particles will undergo a drift of curvature oriented in the east–west direction (and thus west for negatively charged particles). The drift will be at the origin of a current oriented from east to west, called "ring current"[2]. Since the drift velocity is

2 WARNING.– In order to be calculated, the total current in plasma requires the knowledge of the complete trajectory of the particles and not only of their guiding centers (see section 2.2.3.1).

proportional to the mass of the particles, this current is essentially ionic. The ring current generates a magnetic field which opposes the Earth's field at ground level (and which, on the contrary, tends to increase the effective field outside the belts). When the belts are very active (during magnetic storms of solar origin, for example), currents can produce decreases in the magnetic field on the ground of the order of a few hundred nano-tesla.

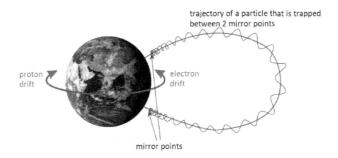

Figure 2.5. *Schematic representation of the trajectory of a particle trapped in the Earth's radiation belts (see section 2.2.3 for the notion of "mirror point"). For a color version of this figure, see www.iste.co.uk/belmont/plasma.zip*

2.2.1.2. Gradient drift

If we consider a null curvature but a magnetic field modulus, which is not constant, and if the gradient of B is sufficiently small $(\nabla_\perp \ll 1/\rho_L)$, then the average force that appears this time is called "mirror force" (F_m). It generates in the same way an average drift given by:

$$\boldsymbol{v}_{dc} = \frac{F_m}{qB} \times \boldsymbol{b}$$

The expression of the mirror force is $\mathbf{F}_m = -\mu \nabla(B)$, with $\mu = \dfrac{mv_\perp^2}{2B}$. Its qualitative interpretation is given in this chapter's Appendices (section 2.4.2).

The existence and the expression of this mirror force can be demonstrated simply in the following approximate way: a charged particle that turns around its guiding center is equivalent to a magnetic dipole. Classically, we know that the magnetic moment of a current loop is equal, in modulus, to $\mu = IS$, which here gives:

$$\mu = \frac{q}{T_c}\pi\rho_L^2 = \frac{q\omega_c}{2\pi}\pi\frac{v_\perp^2}{\omega_c^2} = \frac{1/2\,mv_\perp^2}{B}$$

Vectorially, the magnetic moment of the dipole is opposite to B: we have seen that a positively charged particle rotates in the retrograde (or "left") direction around **B**. Therefore, we must write:

$$\boldsymbol{\mu} = -\mu\, \mathbf{b}$$

The interaction energy between the dipole and the **B**-field is written as $W = -\boldsymbol{\mu}.\mathbf{B} = \mu B$ and the mirror force is that which derives from this energy: $\mathbf{F}_m = -\nabla(\mu B)$, or in other words, $\mathbf{F}_m = -\mu\nabla(B)$, provided that μ is constant, which will be demonstrated later.

COMMENT ON THE NEGATIVE SIGN IN THE EXPRESSION $\boldsymbol{\mu} = -\mu\boldsymbol{b}$.– The cyclotron motion of each particle of the plasma creates a B'-field in the opposite direction to **B**. The particle therefore tends to reduce the field in which it rotates, which is the cause of its gyration motion. It can therefore be expected that, under certain conditions, the plasma is diamagnetic.

2.2.2. Drift in a variable E-field

2.2.2.1. Non-stationary E-field: polarization drift

If, in a constant magnetic field, there is a uniform but non-stationary electric field ($\dot{\mathbf{E}} = \partial_t(\mathbf{E}) \neq 0$), then it can be canceled at any moment by a reference change at the electric drift velocity $\mathbf{v}_{dE}$, but as this velocity is not constant over time, it is necessary to take into account the force of inertia corresponding to this acceleration $\mathbf{F}_p = -m\dfrac{\dot{\mathbf{E}}}{B}\times\mathbf{b}$. The result is an additional drift collinear to $\dot{\mathbf{E}}$:

$$\mathbf{v}_{dp} = \frac{1}{\omega_c}\frac{\dot{\mathbf{E}}_\perp}{B} = \frac{m}{q}\frac{\dot{\mathbf{E}}_\perp}{B^2}$$

called "polarization" drift. The above expression shows that drift velocity is in the opposite direction for the ions and the electrons: it will therefore be at the origin of a volumic current in the plasma, called "polarization current". Since the drift velocity is proportional to the mass (just like the curvature drift), this current will be mainly carried by the ions of the plasma. Its expression is as follows:

$$\mathbf{j}_p \simeq \frac{nM}{B^2}\frac{dE_\perp}{dt}$$

where M is the mass of ions carrying the current. This current is at the origin of a plasma polarization (hence its name) during the passage of an electromagnetic wave in it, and therefore of a large value of the dielectric constant of the plasma in the perpendicular direction in **B**-field.

Figure 2.6 shows the motion of a positively charged particle q in a constant magnetic field directed along the z-axis and an oscillating electric field in time $\mathbf{E} = E_0 \cos(\omega t)\, \mathbf{e}_x$. The amplitude of the field is chosen such that $q\hat{E} = 1$ in the units of the computation, and its frequency such that $\omega/\omega_c = 0.1$.

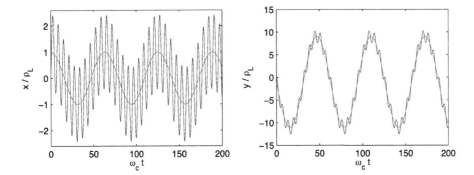

Figure 2.6. *Evolution of the particle position as a function of time under the polarization drift effect. The distances are normalized by the Larmor radius, and the times by ω_c^{-1}. The red curves show the guiding center trajectory. For a color version of this figure, see www.iste.co.uk/belmont/plasma.zip*

Since the particle has no velocity following **z**, the trajectory is entirely in the plane xOy. It has three components: (1) the cyclotron motion (visible in x and y), (2) the crossed fields drift, directed along **y**, and (3) the polarization drift, along **x**. Figure 2.6 illustrates these different components of the motion.

Figure 2.7 shows the same particle launched in the same field, but this time, the electric field pulsation is chosen such that $\omega = \omega_c$ and we see that the nature of the motion is not at all the same. The electric field here resonates with the particle gyration motion (we speak of *cyclotron resonance*). It is therefore important to remember that the decomposition of the motion in terms of drifts is only valid in the approximation of slow variations of physical quantities with respect to the gyro-period of the considered particles (or of small spatial variations over the distance of the Larmor radius).

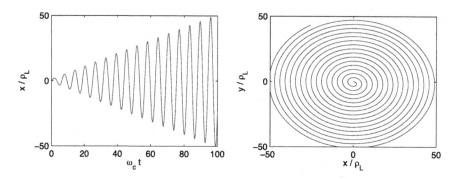

Figure 2.7. *Trajectory of a particle in resonance with the electric field. The electric field work not being zero on average on a rotation, the Larmor radius of the particle linearly increases with the time. We observe a spiral trajectory in the xOy plane*

2.2.2.2. Non-uniform E-field

If it is in space and not in time that the electric field varies, taking into account this variation leads to a modification of the effective electric drift, which is of order 2 in a finite Larmor radius development of the trajectory. If we consider, for example, a one-dimensional variation along x and a Larmor radius ρ_L which is small compared to the scale of the variations of $\mathbf{E}$ with x, then we can perform a Taylor expansion of the trajectory around the middle position in the guiding center (index 0):

$$\mathbf{F} \approx q\mathbf{E}(x = \rho_L \sin \omega_c t) \approx q(\mathbf{E}_0 + \rho_L \sin \omega_c t \mathbf{E'}_0 + \rho_L^2 \sin^2 \omega_c t \mathbf{E''}_0/2 + \ldots)$$

where $\mathbf{E'}_0$ and $\mathbf{E''}_0$ are the first and second derivatives of E with respect to x. We see that this force has a non-zero average value over a period. This is responsible for a modification of the electric drift:

$$\langle \mathbf{F} \rangle \approx q(\mathbf{E}_0 + \rho_L^2 \mathbf{E''}_0/4 + \ldots) \Rightarrow \mathbf{v}^*_{dE} = \frac{\langle \mathbf{F} \rangle}{qB} \times \mathbf{b}$$

This modification does not appear in the most common theories of drifts (which stop at order 1 in ρ_L). It nevertheless has its importance in certain phenomena of instabilities, which are aptly called "drift instabilities" and which are encountered, in particular, in the problems of magnetic confinement (tokamaks). With or without this corrective term, the electric drift velocity obviously remains independent of the electric charge.

2.2.3. *First adiabatic invariant*

We return to the case of a slowly variable magnetic field for which we calculated the drift velocity in section 2.2.1. It is possible to show (the general framework of this demonstration is presented in section 2.3 of this chapter) that the charged particle trajectory satisfies, in this case, certain properties of invariance (or quasi-invariance). An important result is that the following ratio is approximately conserved throughout the trajectory:

$$\boxed{\frac{v_\perp^2}{B} = \text{cste}}$$

The existence of this first approximate invariant, called the "adiabatic" invariant, of the motion constitutes a general result: for "slow" variations, the perpendicular velocity of a particle varies as the square root of the modulus of the magnetic field. This slow variation can be temporal or spatial, explored by the particle along its trajectory.

The following result will be recalled: when the modulus B of the magnetic field increases, the rotation velocity $\omega \approx -\omega_c$ increases as B (in absolute value) while the radius $r \approx \rho_L$ decreases as $B^{-1/2}$ and the perpendicular velocity $v_\perp$ increases as $B^{1/2}$. In the absence of a parallel electric field, the energy conservation of the particles is written as follows: $v_\parallel^2 + v_\perp^2 = \text{cste}$. As B increases, the parallel velocity will therefore decrease.

2.2.3.1. *Magnetic moment and "guiding center" description*

As we have already seen, the magnetic moment associated with the cyclotron motion of a particle around its guiding center has for module $\mu = \frac{mv_\perp^2}{2B}$. The preceding result implies that it is constant during the motion. This bases the "guiding center" description of the particles in a magnetized medium: the particle motion on long timescales with respect to the gyration period can be computed in the same way as a point particle located at the guiding center and provided with a magnetic moment μ anti-parallel to **B**. The existence of this magnetic moment is at the origin of the diamagnetic behavior of a plasma with low frequency.

> NOTE.– The description of the guiding center is simple and fruitful. It nevertheless contains a possible misunderstanding that should be avoided. When calculating a current in a plasma, it must be remembered that the guiding center's motion of ions and electrons is not the only contribution to the current. It is even in many situations negligible compared to the so-called "magnetization current", due to the particles' motion around the guiding center (and therefore related to the magnetic moment).

This fact is illustrated in Figure 2.8: the guiding center's motion is zero, but the density of particles is greater on the left than on the right: it can be seen that there is, indeed, a current that comes from the rotational motions, simply because in the center, there are more particles going down than particles going up.

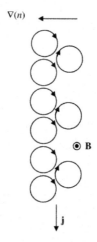

Figure 2.8. *Current balance in a gradient of ionic density without fluid velocity*

2.2.3.2. *Magnetic flux*

If we calculate the magnetic flux through the surface limited by the particle trajectory on a cyclotron rotation, then we find:

$$\phi = BS = B\pi \frac{v_\perp^2}{\omega_c^2} = \pi \frac{m^2}{q^2} \frac{v_\perp^2}{B}$$

This flux, proportional to μ, is therefore also constant during the motion[3]. An important consequence is that in the case of a non-uniform stationary magnetic field explored by the particle along its parallel motion, this shows that the particle has a trajectory that wraps around a magnetic flux tube (in the reference frame moving at the drift velocity), always keeping a radius equal to that of the tube (Figure 2.9).

---

3 *Qualitative interpretation of flow conservation:* Φ is equivalent to the flux of B through a current loop; if it is increased, then an electromotive force is induced in this loop $e_i = -d_t(\Phi)$. The corresponding electric field $\mathbf{E_i}$ would decelerate the perpendicular motion of the particle and thus decrease the Larmor radius; this would decrease the flow that was supposed to increase. This is how Φ can remain constant.

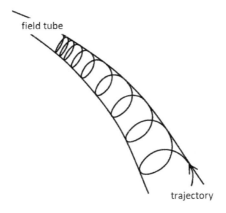

field tube

trajectory

Figure 2.9. *The trajectory wraps around a tube of force of the magnetic field*

2.2.3.3. *Examples*

$\dot{\mathbf{B}} = \partial_t(\mathbf{B})$: purely temporal variation. The conservation of μ constitutes the "betatron" effect: if we increase the magnetic field in a machine, then the perpendicular energy of the particles increases accordingly (particle accelerators).

$\dot{\mathbf{B}} = \mathbf{v}_{//}.\nabla(\mathbf{B})$: spatial variation explored by the particle in its parallel motion. The particle circulates in a "magnetic bottle", in other words, on a convergent tube of force. When it moves towards increasing B, its perpendicular energy increases. In the absence of an electric field or any other force, this is done at constant energy ($v_{\|}^2 + v_{\perp}^2 = cste$): the parallel energy must therefore decrease. Therefore, we have a perpendicular motion that accelerates and a parallel motion that slows down. This transfer of the parallel energy towards the perpendicular energy is done under the effect of the force called "mirror force" (see this chapter's Appendices, section 2.4.2), which is exerted on the magnetic dipole associated with the particle in the guiding center theory. Under the effect of this force, a particle initially heading towards the increasing B slows down, stops at a point called its "mirror point" (where $v_{//} = 0$ and where all the energy is therefore under the form of perpendicular kinetic energy) and goes back in the opposite direction.

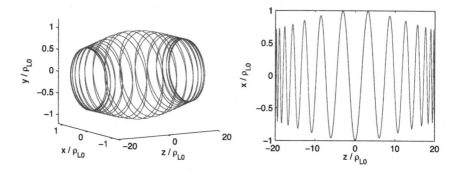

Figure 2.10. *Trajectory of a particle in a magnetic mirror trap*

The phenomena just described are found in many different fields of plasma physics, from magnetic confinement machines for nuclear fusion (see "mirror machines", Figure 2.10) to natural plasmas. In the terrestrial magnetosphere, for example, the particles go back and forth between two mirror points located in each hemisphere somewhat high above the ionosphere; most of them remain "trapped" in the magnetosphere (they also perform a slower drifting motion around the Earth due to the curvature of the lines of force and B-gradients).

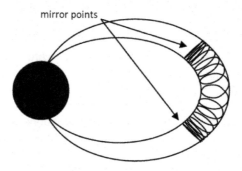

Figure 2.11. *Particle trajectory in the Earth's magnetic field (see Figure 2.5)*

The magnetospheric example is particularly useful to understand the notion of adiabatic invariants. It is briefly described in this chapter's Appendices (section 2.4.3).

2.3. Small disturbances of a periodic motion

We have just seen that when the cyclotron motion in a uniform and stationary field is disturbed by small deviations from uniformity or stationarity, this disturbance modifies its initially periodic trajectory by rendering it "quasi-periodic" and approximately conserving the *adiabatic* invariant μ. This is only a special case of a more general problem: given a mechanical system that has a periodic solution (typically a pendulum), how is the solution modified if the parameters of the system are slowly varied (e.g. the length of the pendulum) compared to the period? We can show that the preceding behavior constitutes a general rule: the solution becomes quasi-periodic and it is possible to characterize the secular evolution by determining an approximate invariant. This generally defines the notion of "adiabatic invariance".

2.3.1. *General results (for a Hamiltonian system)*

A Hamiltonian system is a system without dissipation: the total energy is an exact invariant. Suppose that such a system gives rise to a periodic motion with a period T. We call q the cyclic coordinate of the periodic motion and p_q the conjugate moment of q for the Hamiltonian H of the system. If we slowly disturb this system $(d_f \approx 1/\tau \ll 1/T)$, then we can show, by standard but subtle calculations of Hamiltonian mechanics that we do not present here, that the quantity:

$$J_q = \int_{T_q} p_q \, dq$$

is conserved[4] in first order in T_q/τ.

We also show, in the same order as T_q/τ, that this invariant, called "adiabatic", can also be written in the form of an integral over time:

$$J_q = \int_{T_q} W_q \, dt = <W_q>_{T_q} T_q$$

where W_q is the energy associated with the cyclic coordinate q. This convenient form is often directly usable to quickly find the secular variations of perturbed oscillators: orbit variation of a planet around a star of slowly variable mass, disturbance of a particle trajectory trapped in the potential well of a wave of slowly increasing or decreasing amplitude, and so on.

---

4 This result is written considering a single degree of freedom; in the opposite case, it would be necessary to make a sum on the degrees of freedom.

Regarding the general problem of adiabatic invariants, we shall content ourselves here with the few results that have just been presented without demonstration. However, to fully understand the power of these concepts, let us show their use on a very complete example.

2.3.2. An example with three adiabatic invariants

Among the many magnetic configurations where the notion of adiabatic invariant is useful, we will describe the trajectory of particles trapped in the Earth's magnetic field (see Figures 2.5 and 2.11).

– *Cyclotron motion*: the first periodic motion of a particle is the cyclotron motion. It is slowly disturbed when the particle moves on its field line towards fields that are increasing in strength (by losing altitude) or decreasing in strength (gaining altitude).

$$< W_\perp >_{T_c} = 1/2 \; mv_\perp^2 \text{ and } T_c = 2\pi \; m/qB. ==> J_1 = \frac{2\pi m}{q} \frac{1/2mv_\perp^2}{B}$$

At a numerical coefficient, the J_1 conservation restores the conservation of the magnetic moment μ, which was assumed in section 2.2.3.

– *Bounce motion:* once averaged over the cyclotron period, the particle trajectory (i.e. now the trajectory of its guiding center) describes a back-and-forth motion along the field line, between the two mirror points. It is a new periodic motion, of period T_b; it is slowly disturbed when, during its drift motion, the particle explores lines of force of different lengths. For this new quasi-periodic motion, the preceding formulas give a second adiabatic invariant:

$$J_2 = 1/2m <v_{//}^2> T_b \text{ (T_b being the Bounce period)}$$

CONSEQUENCE.– If a particle is convected (by an electric field) to shorter lines of force (closer mirror points), then the Bounce period decreases and the mean square value of the parallel velocity $<v_{//}^2>$ increases:

$$T_b \text{ decreases} \Rightarrow <v_{//}^2> \text{ grows}$$

This phenomenon of parallel acceleration of a particle trapped between two mirror points that are approaching is called first-order *Fermi acceleration*. This is an effective and important phenomenon in astrophysics (it explains the existence of certain cosmic rays that are among the most energetic particles in the universe).

This phenomenon is often depicted as a ping-pong ball going back and forth between two rackets that are getting closer: rackets provide energy to the ball at each impact.

– *Equatorial drift motion:* once its motion is averaged on the bounce period, the particle (or rather its "magnetic shell") still describes a drift motion around the Earth. For some of these particles, this motion around the Earth is again a periodic motion of period T_d (for instance, the energetic particles in the radiation belts, also called "van Allen belts", at 3 or 4 earth radii). These motions can in turn be disturbed, for example, by variations in the solar wind pressure on the magnetosphere. If these disturbances occur on timescales much longer than the rotation period (which is not obvious in general), then we can still calculate an adiabatic invariant $J_3 = \Phi$, which is the flux of the Earth's magnetic field through the drift trajectory.

Orders of magnitude, for the terrestrial magnetosphere, of the different periodic motions are given in this chapter's Appendices (section 2.4.3). It will be seen that these orders of magnitude are generally sufficiently well separated for each motion to be considered as a slow disturbance of the preceding one. However, this is not always true for disturbances of the last (drift motion), which can be done on timescales of the same order of magnitude as the period (4 h).

In this chapter, we learned how to determine the individual trajectories of particles in realistic imposed fields. In order to actually do plasma physics, it is still necessary to know how to deduce the evolution of the distribution functions (this statistical aspect is the object of the following chapter), how to deduce the moments that are the densities of charge and current and how to solve self-consistently with Maxwell's equations.

2.4. Appendices

2.4.1. *Magnetic field velocity*

We have seen that the existence of the magnetic reference frame R_m does not contribute to the particle trajectories. It comes entirely from Maxwell's equations: a uniform magnetic field is purely magnetic ($\mathbf{E} = \mathbf{0}$) only in a single reference frame R_m; in all other references, it is accompanied by an electric field $\mathbf{E} = -\mathbf{V} \times \mathbf{B}$ related to the velocity $\mathbf{V}_\perp$ of the reference frame considered with respect to R_m. In any reference frame, we can define the "magnetic field velocity" as the velocity at which this "magnetic reference frame" (that is the frame for which $\mathbf{E} = \mathbf{0}$) moves. For stationary and uniform fields, we can say that the magnetic field "moves" as soon as $\mathbf{E} \neq \mathbf{0}$, even if a photo of the field lines at two successive instants strictly gives the same topography of the force lines: in other words, it is necessary to be aware that the electric field "induced" by the movement exists even if $\partial_t(\mathbf{B}) = 0$. This result, which may seem surprising in this form, is, however, commonplace: everyone knows that a piece of wire in a uniform magnetic field is not subject to any electromotor force when it is "immobile", whereas an electromotive force appears as soon as there is a relative motion between the source of the field and the wire: this means that the wire "knows" if the field is "still" or not, even though $\mathbf{B}$ is uniform and stationary.

The apparent paradox of this result only becomes clear if we consider the problem of the sources of the field: how is it that there is a reference point where the charge volume density is zero (for a stationary problem, $\mathbf{E} = 0 \Leftrightarrow \rho = 0$), while this density is non-zero in all of the other reference frames, which have a velocity $\mathbf{V}_\perp$ with respect to this particular reference frame ($\mathbf{E'} \neq 0 \Rightarrow \rho' \neq 0$)? It is true that a body that is neutral in a coordinate system is not generally neutral in others when the system includes currents: from the relativistic formulas of change of reference, $\rho' \neq 0$ as soon as an electrical current $\mathbf{j}$ exists and the reference velocity $\mathbf{V}$ of the change has a component parallel to the current ($\mathbf{j.V} \neq 0$). This effect is fundamentally relativistic (it is based on volume changes), but *it exists at zero order in V/c* (like the one that changes $\mathbf{B}$ to $\mathbf{E}$). Therefore, if we create a dipole magnetic field with a current $\mathbf{j} \neq 0$ flowing in a neutral conductor ($\rho = 0$), we do not create an electric field in this reference frame, but there is an electric field in all other reference frames.

These results, presented for uniform fields, can also extend to non-uniform fields (by considering the system locally). One consequence among others is the "co-rotation electric field" of the magnetized planets. A conductive planet that creates a magnetic dipole is locally neutral in the reference frame that rotates with it. It follows that the dipolar magnetic field it creates "rotates" with the planet (in the preceding sense of the motion of a magnetic field) and that a particle located in space at some distance from there "knows" this movement without intervention of any mechanical collision, simply under the effect of the electric field. This field is then called "co-rotation", since it causes all the particles (more exactly their guiding center) to follow the rotational movement of the planet.

2.4.2. Mirror force

The introduction of the notion of "mirror force" is useful for the intuitive understanding of phenomena. It has been encountered twice in this chapter: on the one hand, to interpret the existence of gradient drift, and on the other hand, to interpret the existence of mirror points.

In the case of a convergent force tube, the interpretation of the mirror force is immediate (Figure 2.12).

The small component δ**B** oriented towards the inside of the tube corresponds to the convergence of this tube. The additional Laplace force associated with this component is downward: it is the mirror force.

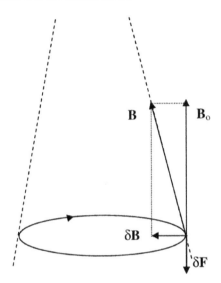

Figure 2.12. *Mirror force δF*

2.4.3. *Movement of particles in the magnetosphere*

The magnetosphere is a prime example for understanding particle motions in a given magnetic field. The following are some guidelines on this example.

2.4.3.1. *Mirror points*

The position of the mirror points is easy to calculate: for a particle of pitch angle α, we have $\sin \alpha = v_\perp / v \propto B^{1/2}$ since the modulus v is constant in the absence of an electric field (constant kinetic energy). We arrive at the mirror point when $\alpha = \pi/2$ ($v_{//} = 0$); if we know the pitch angle and the magnetic field at the equator α_{eq} and B_{eq}, then the mirror point will be where the magnetic field is $B = B_{eq}/\sin^2 \alpha_{eq}$. It will be even lower as the particle leaves the equator with a small angle of attack.

2.4.3.2. *Loss cone*

All particles that have a pitch angle at the equator that is smaller than a certain critical angle α_c are free to reach such low altitudes that they collide with particles in the dense (and thus collisional) part of Earth's high atmosphere; they do not go back up and are therefore "lost". This concerns all particles whose velocity is in the cone of apex angle α_c, which is for this reason called "loss cone" (Figure 2.13). The observations of the magnetospheric distribution functions show that this region of the velocity space is generally empty (Figure 2.14).

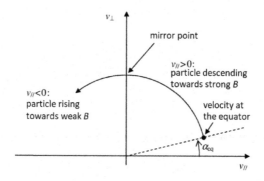

Figure 2.13. *Trajectory of a particle in the plane ($v_{//}$, $v_{\perp}$)*

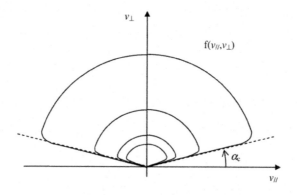

Figure 2.14. *Distribution function with loss cone (iso-contours)*

2.4.3.3. *Timescale orders of magnitude for the terrestrial magnetosphere*

For 50 keV particles on a magnetic field line that intersects the equatorial magnetic plane at 3 earth radii (van Allen belts), we have:

Motion	Electrons	Protons
Cyclotron	$T_c = 50\ \mu s$ (20 kHz)	$T_c = 0.1$ s (10 Hz)
Rebound	$T_b = 1$ s	$T_b = 50$ s
Drift	$T_d = 4$ h	$T_d = 4$ h

Table 2.1. *Typical magnetospheric timescales*

2.4.3.4. *A consequence of the first two invariants: the temperature anisotropy in the magnetosphere*

When we follow the particles coming from the tail of the magnetosphere and moving in the direction of the Earth, their perpendicular and parallel energies vary with respect to the first and second invariants, respectively. This causes the perpendicular energy to increase faster than the parallel energy because:

– $v_\perp^2$ varies as B, that is, approximately as r^{-3} if the field is approximately dipolar;

– $v_\parallel^2$ varies as T_b^{-1}, that is, approximately as L_b^{-2}, or approximately r^{-2} ($L_b \sim v_\parallel T_b$ being the approximate distance between the two mirror points, or the "length of the field line").

We can see that $v_\perp^2$ varies faster than $v_\parallel^2$ (the proof is approximate but the result is correct and can be verified). This difference between the parallel and perpendicular kinetic energies of the particles is entirely in the form of thermal agitation, and a significant anisotropy of the temperatures in the magnetosphere follows: $T_\perp > T_\parallel$.

3

Kinetic Theory of Plasma

The aim of kinetic theory is to find the evolution of the distribution functions of the particles in the plasma with electromagnetic fields which are not imposed, as was done in the preceding chapter for individual particle motion, but which are determined in a self-consistent way from the position and velocity of the particles. In this chapter, we introduce the basic statistical tools that are the kinetic equations.

3.1. Plasma distribution function

The concept of distribution function is introduced in statistical physics. The definition of a plasma does not differ from that of a perfect gas, for example. Nevertheless, the fact that a plasma can be collisionless or strongly dominated by electric and magnetic fields, displays a number of specificities that deserve to be studied. In addition, the kinetic description of plasma is a powerful theoretical and numerical tool for analyzing physical phenomena.

3.1.1. *Definition*

The most complete description of plasma is to consider it as a set of individual particles. To describe the evolution of the system, we apply Newton's second law to each particle. The integration of the differential equations obtained makes it possible to calculate the trajectory of each particle. The implementation of such a description is extremely difficult because, among the forces acting on each particle, the electromagnetic force plays an important role and it depends on the positions and velocities of all the other particles, through charge and current densities. The other complication obviously comes from the large number of particles. We can attempt to directly characterize the set of N particles in the following way: a *particle i* is characterized by a numerical density $\phi_o^{(i)}(t, \mathbf{x}, \mathbf{v})$ defined throughout space. The density $\phi_o^{(i)}$ is a scalar function of time and position in phase space (six-dimensional space of physical positions and velocities); it is defined by the fact that it is equal to 0 everywhere, except where the particle of index i is located, and that its integral on **x** and **v** is equal to 1 (i.e. to the number of particles). We could, for example, model this density function for each variable by a rectangular function of width ε and height $1/\varepsilon$. As far as possible, we shall be satisfied with a description in terms of point particles ($\varepsilon \rightarrow 0$); in this case, the density is not an ordinary function, but it can be expressed with Dirac distributions:

$$\phi_o^{(i)}(t, \chi) = \delta[\mathbf{x} - \mathbf{x}_i(t)] \, \delta[\mathbf{v} - \mathbf{v}_i(t)]$$

We now consider the "population" consisting of a set of N particles. We can again define a numerical density for the population:

$$\phi_N = \prod_{i=1}^{N} \phi_o^{(i)}$$

It is still a numerical density, but in a space with 6N dimensions, that is, in a number N of positions $\chi_1 = \{\mathbf{x}_1, \mathbf{v}_1\}$, $\chi_2 = \{\mathbf{x}_2, \mathbf{v}_2\}$,..., $\chi_N = \{\mathbf{x}_N, \mathbf{v}_N\}$ equal to the *number of particles*. As before, it is normalized (integral equal to 1), but it is non-zero only if the N points χ_i simultaneously correspond to the exact positions of the N particles.

The data of ϕ_N obviously contain much more information than can be manipulated. This information must be reduced by considering the density at one point χ (i.e. in a six-dimensional phase space) due to any of the N particles:

$$\phi_1(t, \chi)$$

It is obtained by integration of the previous density on the N − 1 remaining positions. It is non-zero only if there is a particle in each position χ. However, it no longer expresses itself as a mere product of $\phi_o^{(i)}$: there is a summation over a large number of combinations, since each particle can be found in each of the points.

The summation on all combinations is obvious and simply gives the *sum* of the densities of each of the particles:

$$\phi_1 = \sum_{i=1}^{N} \phi_o^{(i)}$$

Figure 3.1 illustrates what this numerical density, reduced to a function of a position variable, could look like by integrating over velocity. It is an extremely fluctuating quantity, similar to the associated electric field. The field is strengthened when passing near a particle. These very local fluctuations are not useful to understand the plasma's global behavior. It is much simpler and more operational to use a statistical approach and replace an exact description with a probabilistic description.

The distribution function is defined as the probability density in the phase space. It thus determines the average number of particles in a small volume of this space, around a given position and velocity, $f\ (t, x, v)$:

$$\boxed{dn = f(t, x, v)\ d^3v\,d^3x}$$

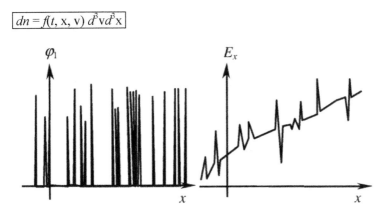

Figure 3.1. *Exact density and the electric field that follows are fluctuating quantities*

The distribution function is a function of seven variables: one for the time, three for position space and three for velocity space. It should be noted that in a plasma, it is generally necessary to distinguish several populations, at least one for positive charges and one for negative charges, and therefore as many distribution functions as populations.

If we want to deduce the density of particles at a point in space from the distribution function, then we integrate over the whole distribution in velocity $n(t, x) = \iiint f(t, x, v)d^3v$, where d^3v is a volume element in the velocity space. We see that there is less information in this distribution function than if we know the exact position of all the particles, but there is much more information than if we only know the particle density at one point in space. Two very different distribution functions can have the same area between the curve and the axis and therefore the same density (Figure 3.2).

The other fluid quantities are deduced from f; they are the moments of order 0, density, order 1, velocity, etc. (see Chapter 1). Figure 3.2 shows a distribution function of a population that has a mean velocity (left) and a distribution function that corresponds to the superposition of a population at rest with a particle beam with a mean velocity.

3.1.2. *Experimental measurement of the distribution function*

The distribution function is not an abstract theoretical concept: if a particle detector is placed on a space probe, what is measured is a velocity histogram that is not very far from a distribution function. Among the instruments for measuring particles in a plasma, mention can be made to Langmuir probes that give access to plasma density and temperature, Faraday-cups that measure current and electrostatic analyzers that measure particle flux at a given energy and in a given direction. The flux measured by an electrostatic analyzer can be easily related to the velocity distribution function and thus, we obtain the distribution function from its measurements (Figure 3.3). We specify the connection below.

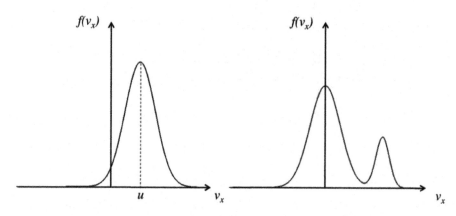

Figure 3.2. *Two distribution functions that give the same density*

Let $J(E, \alpha, \mathbf{x})$ be the flux measured at an energy between E and $E + dE$, in a solid angle $d\Omega$ around the direction of arrival α and at the position of the instrument $\mathbf{x}$. We thus have $J(E, \alpha, \mathbf{x})dE\, d\Omega = vdn$. Here $dn = f\, d^3\mathbf{v} = fv^2 dvd\Omega$ if we make the projection in polar coordinates.

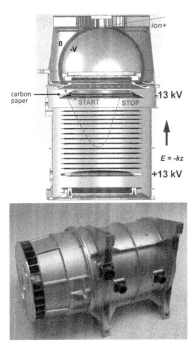

Figure 3.3. *(Top) Diagram of the MSA ion mass spectrometer, which is about to fly onboard BepiColombo. The upper part (hemispherical) analyzes the arrival angle and the energy. The lower part is a time-of-flight analyzer that allows us to determine the m/q ratio of the particle and therefore its nature. (Bottom) Photo of the instrument, in which the cylindrical symmetry and the entrance windows for the particles are seen. For a color version of this figure, see www.iste.co.uk/belmont/plasma.zip*

Since $E = mv^2/2$, we have $dE = mvdv$. Finally, $J = \dfrac{v^2}{m}f$ and we have a very simple relationship between the measured flux and the distribution function. This apparent simplicity hides some technical difficulties. To obtain a distribution function, it is necessary to vary the position (that is, the instrument's position), the energy and the angle of arrival. The variation of energy is obtained by varying the potential difference between the two hemispherical plates that constitute the instrument (see Figure 3.3, top). To vary the angle of arrival, we need as many small windows as angles of arrival, which inevitably limits the angular resolution (Figure 3, bottom). We also see that these angles of arrival are all in one plane; therefore, to access all the possible directions, it is necessary either to use several instruments or to put the analyzer on a rotating platform. This type of instrument is used for the measurement of distribution functions of particles (electrons or ions according to the sign of the potential difference) in space since the end of the 1960s.

3.1.3. *Special case: the Maxwellian distribution*

For a homogeneous and stationary medium, the most known velocity distribution function is the Maxwellian or Maxwell–Boltzmann distribution (written here for a three-dimensional space):

$$f(\mathbf{v}) = \frac{n}{(2\pi)^{3/2} V^3_{th}} e^{-\frac{|\mathbf{v}-\mathbf{u}|^2}{2V^2_{th}}}$$

Its reputation comes from the fact that it represents the distribution function of the velocities in an ideal gas in thermodynamic equilibrium, thanks to the collisions. This form was established by Boltzmann as a hard sphere model, where the atoms of the gas only interact during binary elastic collisions, but it can be generalized to many cases where collisions take place, even plasmas where electrostatic interactions between charged particles are long range. The distribution function is therefore often Maxwellian when collisions are predominant, but it has no reason to be so in a collisionless medium. However, Figure 3.4 shows that in laser-generated laboratory plasmas, the electron distribution function is Maxwellian. On the left-hand side of the figure, the distribution function measured by a probe (by Thomson effect) is plotted at different times during the creation and heating of the plasma by laser absorption, and on the right-hand side, the theoretical expression of several Maxwellian functions with different temperatures is plotted. If we compare the two, we note that the electron distribution function is indeed Maxwellian and that over time the plasma is heated to a temperature of 20 eV.

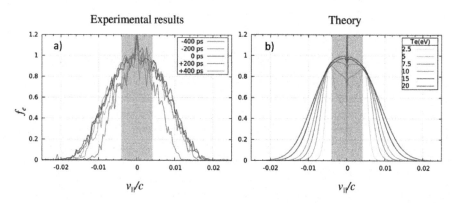

Figure 3.4. *Experimental distribution functions of electrons in a plasma created by laser, in arbitrary units (left), compared with Maxwellian theoretical distribution functions (right). The central part of the figure is not resolved correctly so the variations around 0 are not significant (shaded area). For a color version of this figure, see www.iste.co.uk/belmont/plasma.zip*

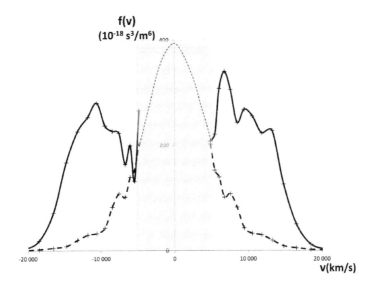

Figure 3.5. *Electron distribution function measured in the Earth's environment by the PEACE instrument onboard the Cluster mission (ESA). The shaded central part is not accessible to measurements. The dotted line represents the perpendicular distribution function, approximately Maxwellian. The solid line represents the distribution function parallel to the magnetic field. Two electron beams of opposite velocities are clearly visible: they are accelerated particles bouncing between two mirror points of the terrestrial magnetic field (source: D. Fontaine)*

The simplest example of a non-Maxwellian distribution function is the so-called bi-Maxwellian distribution function. In a plasma made very anisotropic by a strong magnetic field, the observation of a two-temperature distribution function is not rare, with one temperature in the direction parallel to the field and the other in the direction perpendicular to the field:

$$f(v) = \frac{n}{(2\pi)^{3/2} V_{th//} V_{th\perp}^2} e^{-\frac{(v_{//}-u_{//})^2}{2V_{th//}^2}} e^{-\frac{|v_\perp-u_\perp|^2}{2V_{th\perp}^2}}$$

Another theoretical example of a non-Maxwellian function is shown in Figure 3.2 (right). It can be interpreted as a Maxwellian population plus a particle beam of positive velocity. An experimental example of a non-Maxwellian distribution function is shown in Figure 3.5. This is a measurement made in the Earth's magnetosphere.

3.2. Kinetic equation

3.2.1. *General form*

The kinetic equation is the evolution equation of the distribution function. It is to the distribution function what Newton's second law is to an individual particle. It is deduced from this fundamental law. It has the following general form:

$$\partial_t f + \mathbf{v}.\nabla_x f + <\mathbf{a}>.\nabla_v f = \left(\partial_t f\right)_c$$

The *first member* is universal because it is simply df, the convective derivative of f following the trajectory of an individual particle in phase space. It involves the partial derivative with respect to time $\partial_t f$, the standard spatial gradient (in the position space) $\nabla_x f$, multiplied scalar by the velocity $\mathbf{v}$ and the gradient in the velocity space $\nabla_v f = \dfrac{\partial f}{\partial v_x}\mathbf{e}_x + \dfrac{\partial f}{\partial v_y}\mathbf{e}_y + \dfrac{\partial f}{\partial v_z}\mathbf{e}_z$ scaled by the averaged acceleration <a>. This acceleration is determined by the force $\mathbf{F} = m\mathbf{a}$ acting on an individual particle. In the case where the only forces are electromagnetic, it involves the average value of the fields: $\langle\mathbf{a}\rangle = q/m\left[\langle\mathbf{E}\rangle + \mathbf{v}\times\langle\mathbf{B}\rangle\right]$. The average value here must be understood as a spatial and time average on the same scales δr and δt on which f is defined. We will come back to this point later.

The *second member* contains what we call "collisions". It appears to take into account, in an approximate way, the effects that depend on the microscopic fields other than by their mean values. There are several possible theoretical expressions of this collision term, which means the general kinetic equation is available in different versions, bearing different names, and they are distinguished from each other by the chosen form of the collision term.

Now, let us go back to the notion of average present in the general kinetic equation. Understanding this average is important since the differences between the fields and their mean values are what we can call the "collision fields" or the "microscopic fields". To define the notion of average is therefore equivalent to defining what we call "collisions". We will see that this generalized notion of collisions is very different from the notion of binary collisions that prevails in a neutral gas.

As we have seen, the distribution function $f(\mathbf{v})$ is a probability density concerning the number of particles. Let us specify what this means and what is the statistical ensemble corresponding to this probability. We must first note that to be a smooth function, the distribution f cannot be infinitely precise in $\mathbf{v}$, $\mathbf{r}$ and t. Otherwise, we would find a distribution consisting of extremely sharp peaks (Dirac) corresponding to the positions of each individual particle.

We must therefore define scales: we will say that f corresponds to the number of particles having a velocity in a small volume δv^3 around $\mathbf{v}$, which belong to a small volume δr^3 around $\mathbf{r}$ in real space, the result being averaged over a time $\delta t = \delta r/v$. In this way, f can be a "smooth" quantity, insensitive to the "granular" aspect of the plasma, although it is made of approximately point particles.

The dimension δr characterizing the small volume considered in real space to define f must be neither too large nor too small: it must be between the inter-particle distance (for the statistic to be made on a number of particles large enough for the result to be smooth) and the characteristic scale of the phenomena studied, for example, the Debye length λ_D (to describe these phenomena with a good resolution). This leaves a large flexibility in weakly correlated plasma since, in the solar wind, for example, there is a factor of the order of 10^3 between the two scales d and λ_D, that is, 10^9 for the corresponding volumes.

The direct calculation of the kinetic equation (Klimontovitch 1967, p. 300) does not involve the space and time average of the fields. In fact, it involves the averages of the forces exerted on all the particles found in the small volume of definition. It is not equivalent because this average of forces requires us to know the exact position of all these particles at every moment. This could only be equivalent if all positions were equally probable in the volume δr^3 and in the time interval δt. In this case, the term on the left-hand side of the equation would be sufficient. However, in general, there is a small difference: this is what we must call "collisions" and that is what justifies the presence of the operator of collisions on the right-hand side.

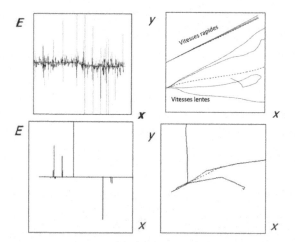

Figure 3.6. *Electric field (left) and particle trajectories (right), for the plasma case (top) and neutral gas (bottom), for different values of the initial velocity. The strongest electric field peaks correspond to passing very near to a particle. The contrast between these peaks and other regions is extremely strong in the neutral gas case, leading to a trajectory consisting of line segments separated by angular points ("binary" collisions). In the plasma case (long-range interaction), the trajectory remains smooth and slowly diffuses with respect to an ideal trajectory (caused by the mean fields only) due to interactions with many distant particles (source: Belmont et al., 2013)*

The collisions term thus comes from the correlations that exist between the position of the particles and the microscopic variations of the fields. If we were interested in very small-scale field fluctuations, that is if δr approached the inter-particular scale, then it would be clear that these correlations would be strong since the field increases considerably when one approaches a particle. However, in a plasma, the effect of these very small-scale peaks remains quite negligible when the δr scale is much larger than d, since their contribution to the correlation integral is proportional to their size. The most important contributions at the end of a collision come from field fluctuations at the largest scales, typically of the order of λ_D. It should be noted that this is radically different from what happens for collisions with neutral particles, where it is the small-scale correlations that are dominant (there are no others). Collisions with the neutral particles are "binary", whereas plasma collisions (for weakly correlated plasmas) are "dependent on many particles". One must therefore be cautious: the word "collision" always intuitively evokes the image of a collision of two hard spheres; this is justified for neutral gas collisions, but not at all in the plasma case. This difference is illustrated in Figure 3.6.

There is no universal expression of the collisional term in the kinetic equation. In general, we try to express this term as a function of the distribution function $f(v)$ and its derivative. The expression of $(\partial_t f)_c$ as a function of f is then obtained via a "collision operator". Several authors have proposed different forms for this operator. We will present some of them in the following sections. All are based on different approximations, more or less refined but often difficult to control. In any case, it is important to remember that, in general, $(\partial_t f)_c$ is not simply a function of f. This is an approximation. It can be shown, however, that this approximation is excellent in the case of collisions between neutral particles, but no equivalent demonstration exists for plasmas. In the neutral case, what simplifies the reasoning is that the only existing correlations are due to very small scales and that they are very strong. To calculate their effect on the large-scale correlation, we can simply add them as so many independent events by invoking the "molecular chaos" (Stosszahlensatz) approximation introduced by Boltzmann.

3.2.2. Vlasov equation

The simplest form of the kinetic equation is the Vlasov equation:

$$\partial_t f + \mathbf{v}.\nabla_x f + <\mathbf{a}>.\nabla_v f = 0$$

It simply consists of considering that the right-hand side $(\partial_t f)_c$ is zero, which is to say that the plasma is collisionless. In this equation, $<\mathbf{a}>$ represents the average acceleration due to the "average" or "collective" fields $\mathbf{E}$ and $\mathbf{B}$; we can write:

$$\partial_t f + \mathbf{v}.\nabla_x f + \frac{q}{m}\left(\mathbf{E} + \mathbf{v}\times\mathbf{B}\right).\nabla_v f = 0$$

Neglecting the right-hand side is equivalent to admitting that the particles do not significantly deviate, on the scales of the studied phenomenon, from their "ideal" trajectory, that is, only due to completely collective fields. The validity of this approximation is therefore relative to the scales studied. If we call "mean free path" λ_{lpm}, the distance over which the trajectory of a particle appreciably deviates from the ideal trajectory, it means that we can study, using the Vlasov equation, all phenomena of scale $L \ll \lambda_{lpm}$. In the solar wind, for example, we can easily study in this way the interaction with a planetary magnetosphere like that of the Earth, because the size of magnetospheres is much lower than the mean free path; however, we cannot study the expansion of the solar wind itself, because its characteristic gradients are generally of the same order of magnitude as λ_{lpm}.

We see in the form of the Vlasov equation that in the absence of collective fields $\mathbf{E}$ and $\mathbf{B}$, any uniform and stationary function is the solution of the equation (a solution that is possibly unstable but a solution anyway). The Maxwellian function plays no particular role in this case, even if it is one of the solutions. In magnetized plasma, that is, if **B** is not zero, we have a similar result: in the absence of an **E** field, any "gyrotropic" function (i.e. in cylindrical coordinates around **B**, independent of the φ coordinate for the velocity variable) is a uniform and stationary solution of the equation. Figure 3.5 shows that the distribution functions observed in the magnetosphere can effectively, significantly deviate from the Maxwellian distribution due to the presence of electron beams. Another example is shown in Figure 3.7. It shows the shape of the electron distribution function in the solar wind: the center looks like a Maxwellian, but there are tails. Suprathermic particles are more important than expected for a Maxwellian.

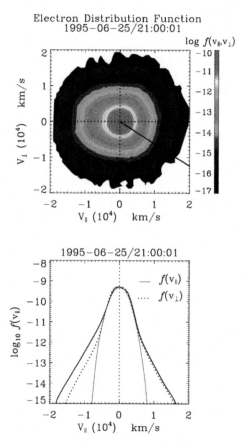

Figure 3.7. *(Top) Electron distribution function measured on board the WIND probe in the solar wind at a distance of one astronomical unit. The figure is centered on the wind's velocity. (Bottom) Projections of the distribution function parallel (solid line) and perpendicular (dashed line) to the magnetic field. We superimposed in blue the outline of a Maxwellian for comparison (source: C. Salem). For a color version of this figure, see www.iste.co.uk/belmont/plasma.zip*

3.3. Different collision operators

3.3.1. *Boltzmann equation*

This is the oldest form of the kinetic equation (1877), by which Boltzmann brought the first kinetic interpretation of hydrodynamics. The thus established collisions operator concerned collisions between neutral particles (binary collisions). It makes two essential assumptions:

– the interactions between particles are all binary and elastic: the field felt at each instant by a particle is either zero or due to a single other particle;

– molecular chaos is present: the different binary collisions thus defined are considered as random and independent events.

The latter hypothesis means that we can ignore the correlations that the collisions themselves could introduce for the position and velocity probabilities of the particles that collide. The collision term is approximated by a hard-sphere model, that is, a zero interaction potential beyond a certain radius and a hard core below. Under these assumptions, the collision term is written as follows:

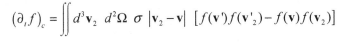

$$\left(\partial_t f\right)_c = \iint d^3\mathbf{v}_2 \ d^2\Omega \ \sigma \ |\mathbf{v}_2 - \mathbf{v}| \ \left[f(\mathbf{v}')f(\mathbf{v}'_2) - f(\mathbf{v})f(\mathbf{v}_2)\right]$$

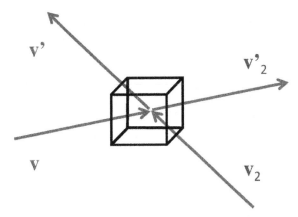

Figure 3.8. *Velocities of two particles that collide before and after elastic collision. The cube represents the volume of the phase space on which the balance of the entries and exits of the Boltzmann collision term is made. For a color version of this figure, see www.iste.co.uk/belmont/plasma.zip*

The velocities before and after the collision are shown in Figure 3.8. The result is a function of $\mathbf{v}$. The coefficient σ has the dimension of a surface, which is called a "differential collision cross-section". It is a function of the relative velocity $v_{rel} = |\mathbf{v}_2 - \mathbf{v}|$ between the two incident particles and of the direction $\mathbf{\Omega}$ of $\mathbf{v}'$ with respect to $\mathbf{v}$. This differential cross-section is such that $f\sigma v_{rel}d^2\Omega$ characterizes the probability per unit of time and per unit of volume for a collision to deviate the incident particle from velocity $\mathbf{v}$ into a solid angle cone $d^2\Omega$ around a given direction $\mathbf{\Omega}$. The distribution function also depends on $\mathbf{x}$ and t, but these variables have not been specified in the formula for simplification purposes.

Let us now show that the Maxwell–Boltzmann distribution function is the stationary solution of the Boltzmann equation. Let us assume that the medium is homogeneous, stationary and not subject to external forces. In this case, all the left-hand-side terms of the equation are null and the collision term must be zero as well. We obtain:

$$f(\mathbf{v}')f(\mathbf{v}'_2) = f(\mathbf{v})f(\mathbf{v}_2)$$

Taking the logarithm of this expression, we find that $\ln f$ must be an invariant during the collision of the two particles, which implies that it is a combination of the invariants already known for an elastic collision, the momentum and the kinetic energy. This leads to a Maxwellian form for $f(\mathbf{v})$.

This seems to distance us a little from plasmas, since we have seen that this interaction of hard-sphere type is not at all representative of the interactions inside a plasma (except in the case of cold plasmas where collisions with neutrals play the most important role).

Nevertheless, it is important to note that the Maxwell–Boltzmann distribution plays a much more general role than the Boltzmann operator. Basically, all operators that assume a superposition of random and independent phenomena (i.e. almost all operators existing in the literature so far) will lead to this distribution. It is in fact a consequence of the "central limit theorem", well-known in statistics. Only operators that interfere with correlations between the different superimposed phenomena could lead to different stationary solutions, possibly closer to what is observed in spatial measurements.

3.3.2. *Landau (1936) and BGL Operators (Lenard 1960, Balescu 1960, Guernsey 1962)*[1]

The case of collisions in a plasma requires more complex calculations than the case of neutral collisions. We must first estimate the microscopic field (assuming the mean field to be equal to zero for simplicity). This field, which in some contexts is called "thermal noise", naturally depends on time and space. It is characterized by its spectrum in the Fourier space $E^2(\omega, \mathbf{k})$. It is necessary, in a second step, to calculate the effect of this field on the individual trajectory of a particle from which we finally deduce the variation of f. For this second step, we first have to focus on the effect of an isolated component $(\omega, \mathbf{k})$ on the trajectory and then sum over all the values of $\mathbf{k}$, the frequency being fixed by a resonant interaction hypothesis $\omega = \mathbf{k}.\mathbf{v}$. The sum over all the values of $\mathbf{k}$ is done simply by calculating the integral over $\mathbf{k}$ of $E^2(\omega, \mathbf{k})$, implicitly assuming a random phase, and therefore the different components $\mathbf{k}$ have no correlation.

This explains why the stationary solution of the kinetic equation, with these two operators, is always a Maxwellian (although the collisions are in no way binary). The sum over $(\omega, \mathbf{k})$, which is reduced to a sum in $\mathbf{k}$ by the resonant hypothesis, is the equivalent of the direct sum on the individual particles in the Boltzmann operator.

In both cases (Landau and BGL), the collision operator gives an equation of the "Fokker–Planck" form, that is:

$$(\partial_t f)_c = -\nabla_v . \mathbf{j}$$

where $\mathbf{j} = \int d^3 k G(v)$ is an integral over $\mathbf{k}$ of a function G, which contains the distribution function $f(\mathbf{v})$, with its reduction in the longitudinal direction (direction of $\mathbf{k}$), $F(v_l) = \int d^2 v_t f(v)$ obtained by integrating over both transverse directions as well as the derivatives of these two functions with respect to the longitudinal component v_l.

As we can see on the "Fokker–Planck" form of the kinetic equation, the collision operator leads, in these cases, to diffusion in the space of velocities. All fine structure in function $f(\mathbf{v})$ will therefore be easily smoothed by collisions. This is a physically interesting result that must be remembered.

---

1 (Landou 1965; Lenard 1960; Balescu 1960; Guernsey 1962).

What distinguishes the two operators (Landau and BGL) concerns the first stage. In estimating thermal noise, Landau simply considers that the causes of this field are individual particles being randomly distributed. As a result, its operator is just the simplified form of the Boltzmann operator that is suitable for the majority of interactions that are the distant interactions, leading to weak deviations. The disadvantage of this approach is that the integral giving the collision operator is then divergent if we do not set a lower limit for k (i.e. for large scales). Landau introduces for this a limit inverse to the Debye length λ_D. It will be seen below (BGL) that on a large scale (small k), it is the (temporal) self-organization of the plasma that modifies the spectrum of the microscopic field and its efficiency to deflect the particles, which can justify this simplification of the integral. Therefore, it is not a static "screening" of the field of each particle as it is often said: such a Debye screening certainly exists at the macroscopic level (an antenna charged in a plasma, for example), but it is wrong to think that this is possible at the microscopic level: we can screen, for example, a sphere of 1 cm positively charged, with a cloud of electrons centered on this sphere, but we cannot screen in the same way each individual ion of the plasma (the same cloud cannot be centered on all the ions at the same time).

The designers of the BGL model also consider that the primary causes of the field are the individual particles. They also take into account the plasma response to any charge deviation. This leads at the end of the calculation to divide the G function of Landau by a factor $|\varepsilon_k|^2$, which makes the spectrum of thermal noise (Figure 3.9) much more realistic and consistent with experimental results. The function ε_k is the dielectric function or "plasma response function" $\varepsilon(\omega, \mathbf{k})$ taken in $\omega = \mathbf{k.v}$. The equation $\varepsilon = 0$ corresponds to the eigenmode (here, the Langmuir mode or "plasma oscillation"). Simplifying the integral to keep only $k\lambda_D > 1$ roughly allows us to avoid the area where the hypothesis of no plasma reaction (that is $\varepsilon_k = 1$) (no reaction of the plasma) is most obviously wrong. However, this also leads to cutting the region where the thermal noise is the strongest (before strongly decreasing for $k\lambda_D > 1$). This can lead to a non-negligible error in the estimation of the collision operator.

To complete the BGL operator, we would have to know how to go beyond the non-self-consistent hypothesis of a lack of correlation between the charge differences, which are at the origin of the field fluctuations. If this hypothesis was well justified in the case of neutral collisions by Boltzmann himself, then this is not the case for collisions of plasma. Currently (and since the 1960s), the BGL operator remains the most complete and scientifically justified operator. Nevertheless, it is extremely difficult to implement, especially in a numerical simulation code, and it is almost never used. The authors often prefer to use the Landau operator, with its cut at the Debye length, despite its rather gross approximation. This operator itself is not simple to implement, and generally, more simple (and completely empirical) operators are used.

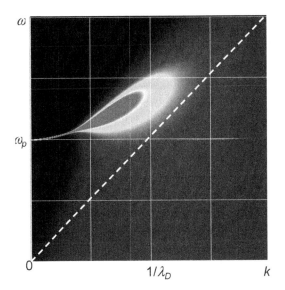

Figure 3.9. *Spectral energy density of thermal noise $E^2(\omega,k)$. The Landau calculus, which assumes $\varepsilon_k = 1$ is essentially valid for $k\lambda_D \gg 1$ and $\omega/\omega_p \ll 1$. Elsewhere, this complete calculation must be used a priori. In particular, we see, for small k, the role of the eigenmode in which the entire density of electrostatic energy is concentrated. For a color version of this figure, see www.iste.co.uk/belmont/ plasma.zip*

3.3.3. *BGK operator (Bhatnagar, Gross and Krook, 1954)*

The BGK operator, also simply called the Krook operator, has the simple form:

$$(\partial_t f)_c = \frac{1}{\tau}[M(f) - f]$$

where $M(f)$ is the Maxwellian distribution, which has the same three first moments as f. This operator has the main qualities of being very simple, to depend only on a single free parameter (τ = collision time = inverse of the collision frequency) and to completely respect the conservation laws for the first three moments. It has another interesting property: the distribution is always approaching a Maxwellian, like all operators of "physical" collisions, as soon as they suppose an infinitely small correlation length for the microscopic field: it is clear that the uniform and stationary solution for this operator is $f = M(f)$.

The main problem with this operator, besides its completely empirical side (we cannot make the Landau or BGL operators approach this form, in any particular case), is that the collision time is unique, whereas all the physical reasoning (based on the previously mentioned complete operators or on much simpler arguments) shows that this time should depend on the velocity as v^{-3}. This has the unfortunate consequence that moments of orders higher than 2, starting with the heat flow, are wrong. Without any particular precaution, this can lead to numerical instability of the codes using such an operator.

There are refinements of the BGK operator that take into account a dependence $\tau(v)$, for example, the Lorentz operator. These operators are always empirical and present a technical difficulty: it is difficult then to make sure that the equation respects the conservation laws of the first three moments (number of particles, momentum and energy).

3.3.4. Fluid consequences of collision operators

We will see in Chapter 4 that fluid equations are obtained by various integrations of kinetic equations with respect to velocity. With a null collision term, the Vlasov equation gives "ideal" equations, while the integration of the different collision terms gives corrective terms with respect to these ideal equations. This determines what are called "transport coefficients": the momentum equations show that there is a corrective term proportional to $\nabla(\mathbf{u})$, the coefficient of proportionality being what is called the "viscosity"; they also show that there is a term proportional to $n_i\mathbf{u}_i - n_e\mathbf{u}_e$, the coefficient being what is called "resistivity" (it is a friction term between ions and electrons). The energy equation shows a term proportional to $\nabla (T)$, the coefficient of proportionality being called "thermal conductivity" and so on.

The most well-known explicit form of the transport coefficients derived from the kinetic theory is the form published by Braginskii (1965). It should be remembered that this form is taken from the Landau operator and not from BGL. These coefficients therefore take into account the effects of the plasma reaction in a very simplified way (cut at the Debye length).

4

Plasma Fluid Modeling and MHD Limit

To solve a problem related to plasma physics, the use of a kinetic equation is often too heavy to be considered, even numerically. We then try to replace this differential equation, which relates the function $f(v)$ and its derivative with respect to v, by a system of algebraic equations, which only relate a small number of quantities. These quantities will be the first moments of the distribution function f, that is, the macroscopic quantities of density, velocity, pressure and so on; they will be called "fluid" quantities. In the same way, we will call "fluid equations" the equations that relate these quantities. We will see that in the general case, it is necessary to introduce a system of fluid equations for each "population" of the plasma. Under some approximations, we will see that this "multi-fluid" system can be reduced to a simpler system, the so-called "mono-fluid": the magneto-hydrodynamic system (MHD).

4.1. Definition of fluid quantities

4.1.1. *Moments of the distribution function*

In a plasma generally composed of electrons and several species of ions, let us consider one "population", that is, as in Chapter 3, any set of identical particles, or at least of the same q/m ratio. The different fluid quantities are defined as the moments of the distribution function.

Density: $n = \int f(v) d^3 v$

Momentum: $nm\mathbf{u} = m \int v f(v) d^3 v$

Kinetic pressure: $\mathbf{p} = m \int (v - u)(v - u)f(v)d^3v$

Heat flux: $\mathbf{Q} = m \int (v - u)(v - u)(v - u)f(v)d^3v$

These different moments have already been presented in Chapter 1, but only in the one-dimensional case. The bold characters indicate tensors of different orders: 1 for $\mathbf{u}$, 2 for $\mathbf{p}$ and 3 for $\mathbf{Q}$. There is a set of moments per population, therefore at least two (electronic and ionic moments).

4.1.2. Definitions of pressure and equation of state

The "kinetic" pressure tensor defined in section 4.1.1 is what we will simply call "pressure" in the remainder of this work, omitting the adjective "kinetic".

First note that this is the "centered" moment $<(v\text{-}u)(v\text{-}u)>$. The pressure therefore only takes into account differences in velocity with respect to the mean velocity (the so-called "fluid" velocity); it therefore measures the velocity fluctuations in a reference frame in which the fluid is at rest (it is therefore a quantity that is independent of the speed of the reference frame).

When we have the opportunity to use the non-centered moment m $<\mathbf{vv}>$, a term called "dynamic" pressure is added: $\mathbf{p}_d = nm\mathbf{uu}$:

$$nm < \mathbf{vv} >= nm < (\mathbf{v} - \mathbf{u})(\mathbf{v} - \mathbf{u}) > + nm\mathbf{uu} = \mathbf{p} + \mathbf{p_d}$$

The tensor $\mathbf{p}$ is also called in other contexts the "stress tensor". Given an elementary surface δS, the vector $\mathbf{p}.\delta S$ represents, in the reference system where $\mathbf{u} = 0$, the volume momentum flow through δS, each particle which crosses the surface carrying its own momentum.

First consider the case of an isotropic distribution function (in the reference system in which its mean velocity is zero): $f(\mathbf{w}) = f(w_x)f(w_y)f(w_z)$ with $f(w_x) = f(w_y) = f(w_z)$. For the sake of simplicity, we have introduced the notation $\mathbf{w} = \mathbf{v} - \mathbf{u}$. The indices x, y and z refer to the components of velocity in a Cartesian coordinate system. According to the definition of the pressure tensor:

$$p_{xy} = \int mw_x w_y f(w_x)f(w_y)f(w_z)dw_x dw_y dw_z = 0$$

Since the average of w_x and w_y is zero by definition, the same reasoning applies to the other non-diagonal terms of the tensor, that is, null. The pressure tensor is therefore diagonal in the isotropic case. We can calculate the value of the diagonal terms:

$$p_{xx} = \int mw_x^2 f(w_x)f(w_y)f(w_z)dw_x dw_y dw_z = nm < w_x^2 >$$

Here again, since the distribution is isotropic, we have $< w_x^2 > = < w_y^2 > = < w_z^2 >$ and we see that the pressure tensor is reduced to a scalar p multiplied by the identity tensor. We usually define *the kinetic temperature T* (in what follows we will refer simply to temperature) as proportional to the root mean square of the velocity: $< w_x^2 > = < w_y^2 > = < w_z^2 > = k_B T/m$. We thus see that this definition induces, in the isotropic case, the state equation connecting density, temperature and scalar pressure: $p = nk_B T$.

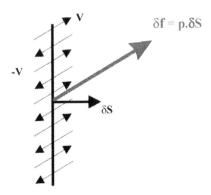

Figure 4.1. *Simple example of non-scalar pressure. If the particles all have one of two velocities ±V (in equal number in each direction), then the distribution function is called "two-beam". The net momentum flux δf is then necessarily carried by the direction common to all the individual momentums and is therefore not collinear with δS. For a color version of this figure, see www.iste.co.uk/belmont/plasma.zip*

In the case of a completely isotropic distribution function, the resulting momentum flux is thus always collinear with the normal vector δS: $\mathbf{p}.\delta S = p \delta S$ (as is the norm in thermodynamics), since the pressure tensor is reduced to a scalar pressure $\mathbf{p} = p\mathbf{I}$. On the contrary, whenever the resulting momentum can have a component orthogonal to δS (i.e. parallel to the surface), the pressure is a complete tensor as in the presence of shear forces or as in the example of Figure 4.1. In a collisional plasma, whose equilibrium state in a homogeneous medium is an isotropic and Maxwellian distribution function, it is the inhomogeneities of the medium that can create anisotropy of the distribution function; this one is then very small, but can play an important role, since it is responsible for the effects of viscosity (it is manifested by the appearance of non-diagonal terms of the so-called "viscosity" in the stress tensor **p**). In a collisionless plasma, on the contrary, the differences in isotropy can be significant and even exist in a homogeneous medium.

In the rest of this chapter, we will mainly consider the case of isotropic media and, therefore, of scalar pressures, in order to reduce the notation and avoid tensorial calculations, which can be heavy without being essential to the physics that we want to describe.

4.1.3. *Kinetic energy*

The kinetic energy of each particle being $1/2mv^2$, it appears as the half-trace of the tensor $m\mathbf{vv}$. The kinetic energy per unit volume is, therefore, the sum of two terms, corresponding to the kinetic and dynamic pressures defined in the previous section, whose expressions are simply found in the isotropic case:

Thermal energy: $E_{th} = \frac{3}{2}p = \frac{3}{2}nk_BT$

Directed energy: $E_d = \frac{1}{2}nmu^2$

It should be noted that the concepts of directed or thermal energy do not have an "absolute" nature, but depend on the manner in which one chooses to separate the set of particles into "populations", each described by its own fluid quantities. The example of the "two-beam" distribution function, shown in Figure 4.2, illustrates this fact: we can choose to describe this system as two mono-kinetic beams (i.e. each of these two populations then having a null temperature) or as a global population that overlaps the two. What was dynamic pressure (and directed energy) in the separated populations becomes kinetic pressure (and thermal energy) in the superposition.

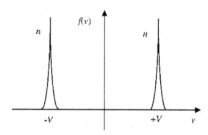

Figure 4.2. *In a collisionless plasma, the dynamic and kinetic pressures of two populations do not add up separately, but the sum of the two is additive. For each beam: $nmu^2 = nmV^2$ and p = 0; for both the beams: $nmu_g^2 = 0$ and $p_g = 2nmV^2$*

In general, neither the kinetic pressure p nor the dynamic pressure ρu^2 are separately additive quantities: only their sum is conserved in the superposition. The additivity of the pressures is, nevertheless, respected in a particular case: it is in the case where all the populations have the same fluid velocity (we will return to it later); this is particularly the case in a collisional environment, which then justifies the use of the notion of "partial pressure".

4.2. Evolution equations of fluid quantities

The kinetic equation governing the evolution of the distribution function of a given population (one or more electron populations, different ionic species) is written, as we saw in Chapter 3, as:

$$\partial_t f + \mathbf{v}.\nabla f + \frac{q}{m}(\mathbf{E} + \mathbf{v} \times \mathbf{B}).\nabla_v f = -\left(\partial_t f\right)_c$$

and the various kinetic equations differ by the expression of the collision term. This term is zero in collisionless plasmas (the Vlasov equation), but is more generally a function of the distribution function $f(t,\mathbf{x},\mathbf{v})$. We name "fluid equations" transport equations (see this chapter's Appendices, section 4.6.1) that connect together the different *moments* of the distribution function with respect to the velocity $\mathbf{v}$. We will show that an *infinite sequence of fluid equations* can be easily derived from the kinetic equation written above.

4.2.1. *Equation of density transport (conservation of the number of particles)*

To obtain the conservation equation of the number of particles, or density transport, it suffices to integrate the kinetic equation with respect to $\mathbf{v}$. The integral of the first two terms immediately provides $\partial_t(n) + \nabla.(n\mathbf{u})$: all you need is to use the integrals of section 4.1.1 defining n and $n\mathbf{u}$, not forgetting that variables $\mathbf{t}$, $\mathbf{x}$ and $\mathbf{v}$ are independent and that the temporal and spatial derivatives can be placed outside of the integrals. The integral of the third term gives identically zero (the electric and magnetic fields do not change the number of particles): for the term in $\mathbf{E}$, it comes, after integration, from the nullity of f for infinite values of the velocity; for the term in $\mathbf{B}$, it comes from the periodicity 2π of the distribution function with respect to the gyrophase[1].

---

1 To see it, we have to write the term $(\mathbf{v} \times \mathbf{B}).\nabla_v f$ in cylindrical coordinates in the velocity space $(v_{\parallel}, v_{\perp}, \varphi)$ and find that it is simply proportional to $\partial f / \partial \phi$. This term is therefore identically zero in the case of a gyrotropic distribution function ($\mathbf{v} \times \mathbf{B}$ and $\nabla_v f$ are orthogonal). More generally, it always gives zero by integrating over ϕ.

A priori, we do not know how to calculate the integral of the collision term until we have actually expressed this term. Nevertheless, it is known in advance that this integral will be identically zero for all types of collisions that retain the number of particles, which is the case as long as there are no chemical reactions (a case that will not be treated in this chapter). This stems from the symmetries that the collision operators must necessarily verify.

The density transport equation, also called the "continuity equation", is therefore written, as we know in hydrodynamics, as:

$$\partial_t n + \nabla.(n\mathbf{u}) = 0$$

4.2.2. Equation of momentum transport

To obtain the equation of momentum transport, it is necessary to multiply the initial kinetic equation by $m\mathbf{v}$ before integrating it (to obtain that of the pressure, it is necessary to multiply the kinetic equation by $m(\mathbf{v} - \mathbf{u})(\mathbf{v} - \mathbf{u})$ before integrating it and so on; we will not detail all of these calculations here). The equation of momentum transport, in the case of a scalar pressure, is written as:

$$\partial_t(nm\mathbf{u}) + \nabla.(nm\mathbf{uu} + p\mathbf{I}) - nq(\mathbf{E} + \mathbf{u} \times \mathbf{B}) = \mathbf{F}_c - \nabla p_c$$

We can rewrite this equation, using the identity $\nabla(nm\mathbf{uu}) = nm\mathbf{u} \cdot \nabla\mathbf{u} + m\mathbf{u}\nabla.(n\mathbf{u})$ and the continuity equation, in the following form usually encountered in hydrodynamics:

$$nm\partial_t\mathbf{u} + nm\,\mathbf{u} \cdot \nabla\mathbf{u} + \nabla p - nq(\mathbf{E} + \mathbf{u} \times \mathbf{B}) = \mathbf{F}_c - \nabla p_c$$

The right-hand side of this transport equation is an integral, for the population considered, of the collision term of the kinetic equation. Let us think a moment on its meaning, because it is one of the ways in which collisions affect the behavior of plasma.

We divided the right-hand side of this equation into two parts: the first term, $\mathbf{F}_c$, represents the frictional forces of the population considered with the other populations, which may, indeed, not be negligible. This is, for example, the case of the friction force with the ions when we consider the movement of the electronic population, and in this case, this introduces the important notion of *resistivity*. The second term, $-\nabla p_c$, represents the forces of "interparticle pressure" internal to the population considered (collisions can change the effective pressure force).

4.2.3. *Transport of energy*

As we have seen, the kinetic energy carried by the particles is the sum of two terms, thermal energy and directed energy. We will derive the equations governing the evolution of these two types of energy and the exchanges between them. For the sake of simplicity, we will neglect the collision terms (which would bring terms of thermal conductivity).

– *Thermal energy*: we obtain the equation of transport of thermal energy by multiplying the kinetic equation by $m(\mathbf{v} - \mathbf{u})(\mathbf{v} - \mathbf{u})$ and by integrating over $\mathbf{v}$. For a scalar pressure (isotropic plasma), we obtain:

$$\partial_t \left(\frac{3}{2} p \right) + \nabla . \left(\frac{5}{2} p \mathbf{u} + \mathbf{q} \right) = \mathbf{u} . \nabla(p)$$

where $\mathbf{q}$ is the heat flux vector[2]. It will be noted that no term coming from the Laplace force appears: this comes from the fact that the Laplace force does not work; it operates exchanges between the momentum in different directions with constant energy.

If, moreover, we can admit that the variations considered are adiabatic (zero divergence of the heat flux: $\nabla . (\mathbf{q}) = 0$), then this equation of transport of the thermal energy can be written in an even simpler (and well-known) form of a relationship between pressure and density:

$$d_t (p / n^{5/3}) = 0$$

The reader can verify this result as an exercise. In this form, we recognize the "equation of state" of adiabatic transformations in thermodynamics (for particles without internal degree of freedom). It is easy to understand why the factor 3 that appears in the 5/3 power stems only from the hypothesis of scalar pressure in a three-dimensional space, that is, from the hypothesis of isotropy of the medium. Note also that this law is certainly not valid generally in non-collisional magnetized plasma, where $\mathbf{B}$ naturally introduces a direction of anisotropy (and where the pressure is not scalar).

---

2 By its definition, we see that the vector $\mathbf{q}$ is defined from the tensor $\mathbf{Q}$ of order 3 by a half-sum on the last two indices: $q_i = \frac{1}{2} \sum_j Q_{ijj}$.

– *Directed energy*: to find the equation of transport of the directed energy, one just needs to multiply scalarly by **v** the equation of momentum transport. For a scalar pressure, neglecting the collisional terms, we obtain:

$$\partial_t\left(nm\frac{u^2}{2}\right)+\nabla.\left(nm\frac{u^2}{2}\mathbf{u}\right)=-\mathbf{u}.\nabla(p)+nq\mathbf{u}.\mathbf{E}$$

– *Total kinetic energy*: the sum of the two preceding equations (thermal energy + directed energy) describes the transport of the total kinetic energy, which is written in the simplest case (with negligible collisions and scalar pressure):

$$\partial_t\left(nm\frac{u^2}{2}+\frac{3}{2}p\right)+\nabla.\left(\mathbf{v}(nm\frac{u^2}{2}+\frac{5}{2}p)+\mathbf{q}\right)=nq\mathbf{u}.\mathbf{E}$$

where $nq\mathbf{u}.\mathbf{E}$, that is the work of the Lorentz forces, describes the transfer of energy from the field to the particles. It can be written $\mathbf{j}.\mathbf{E}$ (see, for example, the Joule effect) by calling the (partial) current due to the population in question.

4.3. Closure equations

4.3.1. *Domain of validity of fluid equations*

The infinite system of fluid equations is rigorously deduced from the kinetic equation and does not suppose any additional approximation: each transport equation is as exact as the kinetic equation from which it was drawn. However, the transport equation of the moment of order 0 (density) involves the moment of order 1 (velocity), the equation of the moment of order 1 involves the moment of order 2 (pressure), the equation of the moment of order 2 involves the moment of order 3 (heat flux) and so on. To have a system containing a finite number of equations, it is necessary to truncate this hierarchy of equations. This truncation implies a hypothesis on the behavior of the last moment introduced into the finite system of fluid equations. This hypothesis is called a closure equation and it involves physical approximations.

The fundamental problem of fluid modeling is therefore that of the closure equation. How to reasonably truncate the infinite system of fluid equations? Since the method is to keep only the first *n* accurate fluid equations (*n* small enough), how to choose the (*n* + 1)th approximate equation, which is supposed to replace the following infinity of fluids equations? The problem arises in very different terms depending on whether the environment is collisional or not. In plasma, which is often a non-collisional medium, it is necessary to make case-by-case approximations and to justify them. When no approximation is valid, we must resign ourselves to reasoning about the complete kinetic equation itself and not about a simpler fluid system.

It must be remembered that fluid equations can only be used in cases where closure equations of the system are justified. The main context in which such conditions are justified is that of "slow variations" in magnetized plasmas (characteristic times much longer than gyro-periods and spatial scales much larger than the Larmor radii, as we will see in sections 4.4 and 4.5).

Analogously, in a non-magnetized plasma, a closure may be used in some cases where space scales are large compared to the Debye length.

We give below examples of closure equations in simple and frequently encountered cases. We limit ourselves to the case, here again the most usual one, where we keep only the first three fluid equations. To find the closure equation is to find an approximate form of the energy transport equation, making an assumption about the expression of the heat flux.

4.3.2. Isotropic pressure

In the simplest case of isotropic plasma, we have seen that the pressure tensor is scalar:

$$\mathbf{p} = \begin{pmatrix} p & 0 & 0 \\ 0 & p & 0 \\ 0 & 0 & p \end{pmatrix}$$

Only one closure equation is necessary. A complete energy equation, with an approximate form of heat flux (Fourier-type law), can provide this closure equation. In this case, the heat flux is a vector that depends on the temperature gradient and is:

$$\mathbf{q} = -\kappa \nabla(T) \text{ (Fourier law: } \kappa = \text{thermal conductivity)}$$

This equation constitutes a natural closure equation, since the third moment (the heat flux) is thus expressed as a function of temperature, that is, the first and second moments (since it is nk_bT). The infinite system of fluid equations is thus limited to only the first three, showing density, velocity and pressure.

We can also often be satisfied with a closure equation of polytropic form:

$$d_t \left(\frac{p}{n^\gamma} \right) = 0,$$

where $p/n^\gamma =$ constant ($\gamma = 1$: isothermal, $\gamma = 5/3$: adiabatic, $\gamma = 3$: one-dimensional adiabatic, $\gamma = \infty$: incompressible)[3]. The choice of the indices γ is most often empirical, it can be experimentally determined. Figure 4.3 shows the measurement of density and temperature in the solar wind over a long distance. It is found that the behavior corresponds to $\gamma \sim 1.3$, which indicates that the behavior of the plasma is neither adiabatic nor isothermal. The solar wind is heated during its expansion.

In non-collisional plasma, the isotropic approximation is generally difficult to rigorously justify from a theoretical point of view. However, experience shows that the anisotropic distribution functions are often unstable and that, even in the absence of collisions, plasma is often not too far from isotropy.

4.3.3. Anisotropic pressure in the presence of a magnetic field

In magnetized plasma, in a frame of reference associated with the magnetic field, a *diagonal* pressure tensor is often considered. This is based on a symmetry hypothesis of the distribution function around **B** ("gyrotropy"). For the same reason, the two diagonal terms corresponding to the directions perpendicular to **B** are assumed to be equal, but they may be different from the parallel pressure term:

$$\mathbf{p} = \begin{pmatrix} p_\perp & 0 & 0 \\ 0 & p_\perp & 0 \\ 0 & 0 & p_{//} \end{pmatrix}$$

Two closure equations are needed in this case. They are often taken in the form:

$$d_t \left(\frac{p_\perp}{nB} \right) = 0 \text{ and } d_t \left(\frac{p_\perp^2 p_{//}}{n^5} \right) = 0$$

---

3 The hypothesis of a polytropic closure equation $d_t(p/n^\gamma) = 0$ does not correspond, for $\gamma \neq 5/3$, to an adiabatic equation, but to a thermal energy equation of the form $\partial_t \left(\frac{1}{\gamma-1} p \right) + \nabla \cdot \left(\frac{\gamma}{\gamma-1} p\mathbf{u} \right) = \mathbf{u}.\nabla(p)$, which does not naturally come, in general, from the initial fluid system.

which cancel the divergence of the heat flux. These laws are sometimes called "double-adiabatic", or "CGL", having been introduced by Chew, Goldberger and Low (1956).

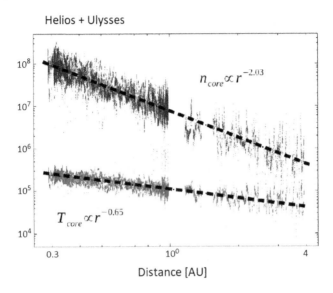

Helios + Ulysses

$$n_{core} \propto r^{-2.03}$$

$$T_{core} \propto r^{-0.65}$$

Distance [AU]

Figure 4.3. *Measurement of the density and temperature of electrons in the solar wind. The Helios (German–American probe) and Ulysses (ESA/NASA) probe measurements are combined to scan a long distance from the Sun. In blue, the density follows an expected r^{-2} law for constant-speed expansion. From the variation of temperature, we can deduce γ (source: S. Stverak). For a color version of this figure, see www.iste.co.uk/belmont/plasma.zip*

In particular, the adiabatic hypothesis is verified whenever disturbances propagate parallel to the magnetic field at high velocities relative to the velocities of all particles. In the opposite hypothesis (slow propagation), it is rather isothermal laws (polytropic laws with $\gamma = 1$) which are suitable.

It is also possible, like in the isotropic case, to use empirical polytropic laws for the components of the pressure parallel and perpendicular to the magnetic field:

$$d_t \left(\frac{p_\perp}{n^{\gamma_\perp}} \right) = 0 \text{ and } d_t \left(\frac{p_\parallel}{n^{\gamma_\parallel}} \right) = 0$$

4.3.4. *General case*

In a collisional medium, the binary interactions always result in a fluid behavior and, therefore, in the existence of a closure equation. In collisionless plasma, only the collective fields can possibly provide such behavior.

When no closure law is satisfactory, it is necessary to work directly on the distribution function and thus with the Vlasov equation (kinetic treatment; see Chapters 3 and 6). This means that the system of equations for the first three moments is not a closed system and that the evolution of these three variables cannot be determined without knowing more information. In this case, two populations that initially have the same first three moments may then have different evolutions and moments that become different. This is shown in Figure 4.4: two populations are shown whose distribution functions have substantially the same density, same mean velocity and the same perpendicular and parallel thermal velocities. It is easy to imagine that the evolution of these distribution functions may be different and that their moments, a moment later, may take on different values.

4.4. Multi-fluid description of plasma

4.4.1. *Concept of "population" and multi-fluid system*

In order to achieve a "fluid" description of plasma and thus replace the distribution function of all particles by only a few macroscopic parameters (density, velocity, pressure), we must first group the particles into the smallest possible number of "populations". Among the plasma particles, it is necessary to distinguish at least the different "species" present, ions or electrons, characterized by their charge and mass, or more precisely by their q/m ratio. However, even within each species, it may be necessary to distinguish several subsets of particles, for example, depending on their origin[4], the main criterion always being that each of the

---

4 In magnetospheric physics, for example, it is possible to imagine that a detector on a satellite measures an electronic distribution function including two peaks, one consisting of low-energy particles ("cold" electrons), coming from an ionospheric source located on the same line of force, and the other consisting of higher-energy particles ("hot" electrons), convected perpendicular to **B** from the magnetospheric tail. It is likely that these two populations of electrons behave differently with respect to field variations.

"population" thus identified actually respects a system of fluid equations, with a justified closure equation.

Each population, hereafter indicated by the index j, is characterized by its first moments n_j, $\mathbf{u}_j$, p_j. For the sake of simplicity, let us first consider a fluid case called "ideal", that is, with the second members due to the collisions equal to zero (this would not be the case, for example, if electron–ion collisions caused resistivity). We will also assume that the pressures are scalar; an anisotropic fluid theory would be constructed quite similarly, but many results are not critically dependent on this simplifying assumption.

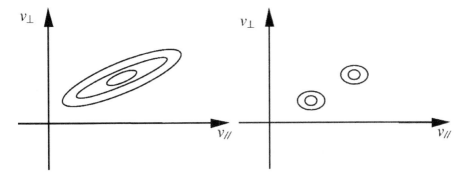

Figure 4.4. *Iso-contours of distribution functions that initially have their first three moments identical and which will not necessarily have the same subsequent evolution, as their moments may then take on different values*

Suppose a system with N populations. For each population, we have five equations (two scalar equations and one vector) for five fluid variables [n_j, $\mathbf{u}_j$, p_j] (two scalar variables and one vector), omitting the heat flux $\mathbf{q}_j$ (which we will assume from the beginning of the next section to be connected, by a closure equation, to the moments of lower orders). The system also covers six $\mathbf{E}$ and $\mathbf{B}$ field variables. Maxwell's equations provide six additional independent equations. We thus have, in total, a system of *$5N + 6$ equations with $5N + 6$ unknowns:*

$$\partial_t(n_j) + \nabla.(n_j\mathbf{u}_j) = 0 \qquad (N)$$

$$\partial_t(n_j m_j \mathbf{u}_j) + \nabla.(n_j m_j \mathbf{u}_j \mathbf{u}_j + p_j) = n_j q_j(\mathbf{E} + \mathbf{u}_j \times \mathbf{B}) \qquad (3N)$$

$$\partial_t\left(n_j m_j \frac{u_j^2}{2} + \frac{3}{2}p_j\right) + \nabla.\left[\mathbf{u}_j\left(n_j m_j \frac{u_j^2}{2} + \frac{5}{2}p_j\right) + \mathbf{q}_j\right] = n_j q_j \mathbf{u}_j.\mathbf{E} \qquad (N)$$

$$\nabla \times (\mathbf{E}) = -\partial_t(\mathbf{B}) \qquad (2)$$

$$\nabla \times (\mathbf{B}) = \mu_0 \mathbf{j} + \frac{1}{c^2}\partial_t(\mathbf{E}) \quad \text{with} \quad \mathbf{j} = \sum_j n_j q_j u_j \qquad (2)$$

$$\nabla.(\mathbf{E}) = \frac{\rho}{\varepsilon_0} \qquad \text{with} \quad \rho = \sum_j n_j q_j \qquad (1)$$

$$\nabla.(\mathbf{B}) = 0 \qquad (1)$$

* in parentheses is the number of scalar equations corresponding to each line.

According to the series of equations above, the system is not explicitly closed. For each population, the last equation of the fluid part of this system is the complete energy equation, in which the "heat flux" variable $\mathbf{q}_j$ has not been expressed. The choice of the closure equation will determine the actual form of this last equation. Depending on the case, any of the closure equations presented in the previous section may be used; practically, a very simple equation of the type $p_j/n_j^\gamma = \mathrm{cst}$ is often used. In this chapter, we will choose an adiabatic hypothesis, that is, we will keep the energy equation above, with:

$$\nabla.(\mathbf{q}_j) = 0$$

It is important to note that each population can, in principle, have a different closure equation[5].

4.4.2. Slow variation hypothesis

In the study of individual trajectories carried out in Chapter 2, a first "slow variation" hypothesis has been mentioned: the first adiabatic invariant, as well as the notion of drift velocity could be introduced by considering slow perturbations with respect to the period of the cyclotron movement. What we will now call "slow variation hypotheses" includes this same condition on the perpendicular motion, as

5 We can, for example, have adiabatic ions and isothermal electrons.

well as a second condition on the parallel motion of the electrons, which brings an additional simplification, called quasi-neutrality, of the system.

We will see that this slow variation hypothesis, justified when we study a system on "large" spatial-temporal scales (see section 4.4.3), makes it possible to considerably simplify the system of fluid equations to be solved. Indeed, without the approximations of "slow variations", the above system is generally quite heavy: it is a 16×16 system (16 equations and 16 unknowns) in the minimal case, where there are only electrons and a single species of ions; 21×21 as soon as there are two species of ions to distinguish and so on.

4.4.3. Statement of slow variation hypotheses

Let us consider the low-frequency limit and the large-scale limit, that is, $\partial_t \approx 1/\tau << \{\omega_{cj}, \omega_{pe}\}$ and $\partial_x \approx 1/L << \{1/\rho_{Lj}, 1/\lambda_{De}\}$. This means that the characteristic time τ of the studied phenomena is long, on the one hand, with respect to the gyration time of all the particles around the magnetic field $(1/\omega_{cj})$ and, on the other hand, with respect to the return time to the quasi-neutrality $(1/\omega_{pe})$ related to the parallel motion of the electrons. Similarly, the spatial scale of the phenomena must be large relative to the Larmor radius of all particles perpendicular to the magnetic field and large relative to the electron Debye length along **B**.

4.4.4. Consequences

To study the consequences of the slow variation hypotheses that have just been defined, we need to use the system of equations in section 4.4.1 to make the characteristic scales τ and L of the studied phenomena tend to infinity, that is, the derivatives ∂_t and ∂_x tend to zero. We immediately see that:

$$\mathbf{E} + \mathbf{u_j} \times \mathbf{B} \rightarrow 0$$
$$\rho \rightarrow 0$$
$$\mathbf{j} \rightarrow 0$$

CONSEQUENCE 1: FROZEN FIELD.– The first equation is extremely important; it means that the different populations all have substantially the same perpendicular fluid velocity:

$$\boxed{\mathbf{E} = -\mathbf{u_j} \times \mathbf{B} \Leftrightarrow \mathbf{u}_{j\perp} = \frac{\mathbf{E}}{B} \times \mathbf{b}}$$

This velocity is, in fact, the mean velocity (drift velocity) of each individual particle in the case of slow variations (in relation to $1/\omega_{cj}$ and ρ_{Lj}). This common velocity is the first cause of the collective ("fluid") behavior of particles with respect to *perpendicular* motions. This common movement of all populations at the "velocity of the magnetic field" is found in all the slow variation theories, especially MHD, which we will study a little further. We speak of the "frozen field" in the plasma (or frozen plasma in the field). Section 4.6.2, in the appendices of this chapter, develops an important consequence of this law: the identity of the field lines during the movement (conservation of "connections") and the problem of magnetic "reconnection". Two more immediate consequences can already be retained:

– at the zeroth order, $E_{//} \approx 0$ (i.e. $E_{\parallel} \ll E_{\perp}$). If the field is electrostatic in nature, then the magnetic field lines become quasi-equipotential;

– for non-relativistic plasmas, we have $u_{\perp} = E_{\perp}/B \ll c$. This implies that the electrical energy is always negligible compared to the magnetic energy: $\varepsilon_0 E^2 \ll B^2/\mu_0$, and also that the displacement current $\varepsilon_0 \partial_t \mathbf{E}$ is negligible compared to the conduction current $\mathbf{j}$ carried by the particles.

CONSEQUENCE 2: QUASI-NEUTRALITY.– The second equation constitutes the so-called quasi-neutrality approximation associated with slow variations (with respect to $1/\omega_{pe}$ and λ_{De}). It does not mean that the space charge is rigorously zero, but that it is much lower than the charge densities at $n_j q_j$ due to each individual population: it is of order 1 in the development in slow variations, such as $\nabla.(\mathbf{E})$, while the $n_j q_j$ charges are of order 0.

CONSEQUENCE 3: QUASI-NULL CURRENT.– In the same way, the third equation shows that the total current is of order 1 as well, that is, very much less than the currents $n_j q_j \mathbf{u_j}$ of each population (which are of order 0). This does not prevent currents from playing an important role in plasma dynamics.

The above two equations (quasi-neutrality and quasi-null currents) are due to the collective behavior of the different low-frequency plasma populations for *parallel* movements to the magnetic field.

Using all the approximations that have just been presented, the previous system becomes lighter. By sticking to the lowest order in the development in "slow variations" (with respect to ∂_t and ∂_x), equation $\mathbf{E} = - \mathbf{u_j} \times \mathbf{B}$ makes it possible to eliminate $2N - 1$ fluid equations and $2N - 1$ variables. The reason why we can only eliminate $2N - 1$ equations and not $2N$ equations is that these equations are not linearly independent at this order: on summing up the terms $n_j q_j (\mathbf{E} + \mathbf{u} \times \mathbf{B})$, we obtain $\rho \mathbf{E} + \mathbf{j} \times \mathbf{B}$, which is a term of order 1 (and therefore null at order 0). We must, therefore, keep in the system this equation of order 1, which is the sum over the populations j of all the momentum equations. We have thus reduced the dimension $5N +$ 6 to a dimension $3N + 7$. Moreover, two Maxwell equations are simply replaced by $\rho \approx 0$ and $\mathbf{j} \approx 0$, which is substantially simpler, and the displacement current has faded away. For this system to approach the classical system of fluid mechanics, we only have to reduce as much as possible the value of the number N of populations.

4.5. MHD (magneto-hydrodynamic) description

4.5.1. *Reduction of the number of populations for the same species*

Within the same species (same q/m ratio), we can always try to group several populations, because the grouping of two populations is a new population. All the "exact" equations of the fluid system then keep the same form as the equations of the elementary populations, with the moments of the global population g being simply connected to those of the elementary populations by:

$$n_g = \Sigma n_j, \ n_g m_g u_g = \Sigma n_j m_j u_j, \ \mathbf{p}_g + n_g m_g u_g u_g = \Sigma \left(\mathbf{p}_j + n_j m_j u_j u_j \right)$$

$$n_g m_g = \Sigma n_j m_j, \ \rho_g = \Sigma n_j q_j, \ \mathbf{j}_g = \Sigma n_j q_j u_j$$

It should be noted that the kinetic pressures only add up in cases where the different populations are described by the same velocity field $\mathbf{u}$ (this is the condition under which we can define the concept of partial pressure, encountered in thermodynamics).

However, the fluid system does not solely consist of exact equations: the closure equation is an approximation and may be different from one population to another. It limits the possibilities of regrouping:

– the existence of elementary closure equations does not generally lead to the existence of a global closure equation;

– even though an identical closure equation is valid for all populations, it is not automatically valid for the global population as soon as this equation is nonlinear. If we consider, for example, polytropic closure laws, then it is not obvious, in general, that $p_1 / n_1^\gamma = k_1$ and $p_2 / n_2^\gamma = k_2$ necessarily entail $p_g / n_g^\gamma = k_g$.

4.5.2. *Case of N = 2: bi-fluid system and MHD*

The general bi-fluid system. The simplest case that we can solve is that where there are only two species: the electrons and a single species of ions (positive). If it has been possible to group all the electrons into one population and the same for the ions, then we are dealing with a "bi-fluid" system. This can occur under a variety of conditions, and we will assume here that both species of particles are *adiabatic*[6]. The bi-fluid system is generally a system of dimension $5N + 6 = 16$.

MHD. In the case of slow variations, the dimension of the bi-fluid system decreases and, according to the general rule derived in the previous part, it should become $3N + 7 = 13$. Let us show that the system is, in fact, only a dimension eight.

The quasi-neutrality equation requires $n_+ \, q_+ + n_- \, q_- \approx 0$, that is:

$$n_+ \approx n_- \approx n$$

in the case where there are no multi-charged ions[7].

In the same way, the equation for the current requires that $n_+ \, q_+ \, \mathbf{u}_+ + n_- q_- \mathbf{u}_- \approx 0$, and thus, since $n_+ \, q_+ + n_- q_- \approx 0$:

$$\mathbf{u}_+ \approx \mathbf{u}_- \approx \mathbf{u}$$

It can thus be seen that a bi-fluid system, in the case of slow variations, is in fact characterized only by a single fluid velocity and a single density, the velocity and the density of the positive particles being equal at the zeroth order to those of negative charges (or proportional to them in the case of multi-charged ions).

The equations governing the evolution of the two populations' momentum only differ by the masses and the pressure terms. We can thus obtain a system of very simplified equations by taking the sum of conservation equations for the ionic species and the electronic species. The sum of the Lorentz forces $\sum n_j q_j \mathbf{u} \times \mathbf{B}$ provides the Laplace force on the plasma $\mathbf{j} \times \mathbf{B}$; the sum of the electric forces $\sum n_j q_j \mathbf{E}$ is equal to the global Coulomb force $\rho_g \mathbf{E}$ acting on the plasma, but this term is generally not considered; it is, indeed, negligible compared to the Laplace force as we consider the usual non-relativistic hypothesis (electrical energy << magnetic energy; displacement current << conduction current).

---

6 This is verified if the parallel thermal velocities of the two species are much smaller than the ratio of the characteristic scales $L_{//}/\tau$ of the studied phenomena.

7 In the opposite case, we would have $Z^+ n^+ = Z n^- \approx n$ by assuming $q_+ = Z_+ \, e$ and $q_- = -Z_- e$.

The bi-fluid system, under the assumption of slow variations, thus becomes an apparently mono-fluid system (a single density; a single velocity). It is called a magneto-hydrodynamic system, or MHD, and will be retained as:

$$
\begin{cases}
\partial_t n + \mathbf{V}.\,(n\mathbf{u}) = 0 & [4.1] \\[2mm]
nm\partial_t \mathbf{u} + nm\mathbf{u}.\,\mathbf{V}\mathbf{u} = -\mathbf{V}p + \mathbf{j} \times \mathbf{B} & [4.2] \\[2mm]
\partial_t \mathbf{B} = -\mathbf{V} \times \mathbf{E} & [4.3] \\[2mm]
d_t\left(\dfrac{p}{n^\gamma}\right) = 0 & [4.4]
\end{cases}
$$

$$
\begin{cases}
\mathbf{E} = -\mathbf{u} \times \mathbf{B} & [4.5] \\[2mm]
\mathbf{j} = \mathbf{V} \times \mathbf{B}/\mu_0 & [4.6] \\[2mm]
\mathbf{V}.\,\mathbf{B} = 0 & [4.7]
\end{cases}
$$

In this system, the first four equations contain time derivatives. Therefore they make it possible to calculate the evolution over time of the density, velocity, magnetic field and pressure. The following three equations connect these different quantities at the same time t.

The first equation is the conservation of the number of ions or electrons. It is valid in most cases, in the hypothesis of absence of electron–ion recombination.

The second equation describes the conservation of the momentum of the quasi-neutral fluid consisting of electrons and ions. The mass m is defined as the sum of the electronic and ionic masses, $m = m_e + m_i \simeq m_i$, and the pressure as the sum of the electronic and ionic pressures, $p = p_e + p_i \simeq nk_B(T_e + T_i)$. Collisions having been neglected from the beginning of the derivation started in section 4.4 (the system is called "ideal" MHD), no viscosity force appears in the right-hand side. Since the fluid is assumed isotropic, the pressure is scalar. This equation could be generalized by introducing a pressure tensor, potentially anisotropic, and a viscosity term modeling, as in the Navier–Stokes equation, the diffusion of the momentum occurring under the effect of collisions.

The third equation, the Maxwell–Faraday equation, is obviously valid regardless of any hypothesis. Given Ohm's law [4.5], this equation directly links the evolution of the magnetic field to the shape of the fluid velocity field, in the form $\partial_t \mathbf{B} = \mathbf{\nabla} \times (\mathbf{u} \times \mathbf{B})$.

Equation [4.4] is the closure equation, which is, as we have seen earlier, an approximation. The equation given here corresponds to a simple closure called "polytropic". We often take $\gamma = 5/3$, which would correspond to an adiabatic hypothesis in the absence of interaction with the electromagnetic field. We could use an even simpler form, $p = Kn^\gamma$, which is equivalent to the previous form if the value of p/n^γ is the same everywhere in the initial condition. In this case, equation [4.4] is no longer an evolution equation (no more time derivative), and the knowledge of n is sufficient to ensure the knowledge of p at each instant. On the contrary, we can use a more complete energy equation:

$$\partial_t \left(nm\frac{u^2}{2} + \frac{3}{2}p \right) + \mathbf{\nabla}.\left(\mathbf{u} \left[nm\frac{u^2}{2} + \frac{5}{2}p \right] + \mathbf{q} \right) = \mathbf{j}.\,\mathbf{E}$$

where $\mathbf{q} = 0$ (complete adiabatic hypothesis) or $\mathbf{q} = -\kappa\nabla T$ (thermal conductivity due to collisions) may be encountered, depending on the case.

Equation [4.5] is the "ideal Ohm's law". It is an approximation related to the hypotheses of slow variations and to a hypothesis of zero resistivity (few collisions), but it can be replaced, when necessary, by less simplified equations (generalized Ohm's laws).

Equation [4.6] is the Maxwell–Ampere equation. The displacement current has been neglected, which is the valid approximation for the slow variations considered.

The last equation, finally, expresses the nullity of $\mathbf{\nabla}.\mathbf{B}$ and is always valid.

The MHD system is a *mono-fluid* plasma description, since it only shows a single density and a single velocity, but it must be remembered that it is in fact the system describing one or the other of the two fluids (positive or negative) of a bi-fluid system, because in the particular case of slow variations, the two fluids show the same behavior. Density, in particular, is not the sum of densities $n_+ + n_- = 2n$ (as would be the case if it were a grouping of populations), but simply $n_+ = n_- = n$ (for mono-charged ions).

Note that beyond the case of plasma, the MHD system is adapted to the description of the dynamics of any conductive fluid (in which there is a free electron population).

4.5.3. *The Laplace force in MHD: magnetic pressure and stress*

The MHD system we have just derived is identical to that of fluid mechanics, except that a Laplace force term appears in the momentum equation and two Maxwell equations must be added to deal with the evolution of variable $\mathbf{B}$, which thus appeared.

Since the current density is directly related to the magnetic field by Ampere's law, the Laplace force is entirely determined by the shape of the $\mathbf{B}$ field:

$$\mathbf{j} \times \mathbf{B} = \frac{1}{\mu_0} (\nabla \times \mathbf{B}) \times \mathbf{B} = -\nabla \frac{B^2}{2\mu_0} + \frac{1}{\mu_0} \mathbf{B}.\nabla \mathbf{B} = F_P + F_T$$

We see that the Laplace force (per unit volume) is written as the sum of two components. The first value $F_P = -\nabla B^2 / 2\mu_0$, which is therefore equal to the opposite of the magnetic pressure gradient (or magnetic energy density). It behaves like a pressure force, which will tend to push the plasma from areas of strong magnetic field to areas of weaker field. We speak of magnetic pressure force (see illustration in Figure 4.5).

The second component is related to the local curvature of the $\mathbf{B}$ field. It can be expressed more intuitively by introducing the Frenet coordinate system $(\mathbf{b}, \mathbf{n})$ associated with the field line, and the curvilinear abscissa s along this line. We can write the stress in the form:

$$F_T = \frac{d}{ds}\left(\frac{B^2}{2\mu_0}\right) \mathbf{b} + \frac{B^2}{\mu_0 R} \mathbf{n}$$

where $\mathbf{R}$ is the local radius of curvature of the field line. It can be seen that this stress force cancels at all points (via its first term) the component parallel to the field of the magnetic pressure. This is expected since the Laplace force $\mathbf{j} \times \mathbf{B}$ has, of course, no component in the direction of $\mathbf{B}$. The remaining part, perpendicular to the field, is inversely proportional to the radius of curvature of the field line: the more the line is curved, the greater the force exerted by the field on the plasma (and dragging it towards the center of curvature) is important.

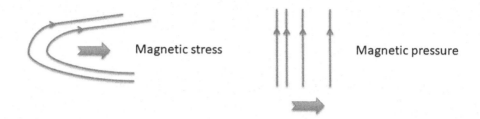

Figure 4.5. *Oriented blue lines represent magnetic field lines. (Left) Curved lines and the direction of the associated magnetic stress force. (Right) A gradient of the modulus of the B field perpendicular to the field lines and the direction of the magnetic pressure force*

The dimensionless parameter β had previously been defined as the ratio of the thermal pressure p of the plasma to the magnetic pressure:

$$\beta = \frac{nk_BT}{B^2/2\mu_0}$$

The value of this parameter gives an indication of the effects that dominate plasma dynamics. When $\beta \ll 1$, the effects related to the Laplace force dominate, and the topology of the magnetic field, imposed by its boundary conditions, will determine the spatial distribution of the plasma and its evolution. Conversely, when $\beta \gg 1$, the thermal pressure dominates. The dynamics of the plasma then approaches that of a neutral fluid, and the shape of the magnetic field lines is imposed (via Ohm's law) by the movement of the fluid.

Figure 4.6. *Coronal loops (source: image by TRACE satellite [Transition Region and Coronal Explorer], NASA). For a color version of this figure, see www.iste.co.uk/belmont/plasma.zip*

The plasma of the solar corona provides an example of a small β environment, in which the magnetic field is intense ($B \sim 100$ mT) and the particle density is quite low ($n \sim 10^9 \text{cm}^{-3}$). The parameter β is then of the order of $\beta \sim 10^{-6}$. It can be seen in Figure 4.6 that the spatial distribution of the plasma seems to be totally determined by the configuration of the magnetic field loops of the solar surface.

4.5.4. Ohm's law and the effect of non-ideality

In Faraday's equation of the MHD system, the electric field has been eliminated by using $\mathbf{E} = -\mathbf{u} \times \mathbf{B}$. We must remember, however, that, for simplicity, it has been assumed since the beginning of this chapter (see section 4.4.1) that the fluid is "ideal" and that any effect of collisions is negligible. In collisional plasma, it is often necessary to take into account the terms of electron–ion friction, and we obtain:

$$\mathbf{E} = -\mathbf{u} \times \mathbf{B} + \frac{\mathbf{j}}{\sigma}$$

where the conductivity σ of the plasma appears. This equation is called Ohm's law. The term $-\mathbf{u} \times \mathbf{B}$ is added to the usual form used by electricians as soon as there is a magnetic field and a movement with respect to this field. Since the current density $\mathbf{j}$ is related to the magnetic field by Ampere's law, it can be noted that the presence of a non-zero resistivity (of non-infinite conductivity) will result in the addition of a diffusive term in the equation of evolution of the magnetic field; Faraday's equation indeed becomes:

$$\partial_t \mathbf{B} = \nabla \times (\mathbf{u} \times \mathbf{B} - \nabla \times \mathbf{B}/\mu_0\sigma) = \nabla \times (\mathbf{u} \times \mathbf{B}) + \Delta\mathbf{B}/\mu_0\sigma$$

where Δ is the Laplacian vectorial operator. The second term on the right-hand side describes the diffusion of the magnetic field under the effect of the resistivity, with a diffusion coefficient $D_m = 1/\mu_0\sigma$. In a system characterized by a typical spatial dimension L and a typical velocity U, the order of magnitude of the terms on the right-hand side can be evaluated as $\nabla \times (\mathbf{u} \times \mathbf{B}) \sim UB/L$ (convective term) and $\Delta\mathbf{B}/\mu_0\sigma \sim B/L^2\mu_0\sigma$ (diffusive term). To quantify the respective effect of diffusion and convection in the evolution of the magnetic field, we often introduce, by analogy with the Navier–Stokes equation, a dimensionless parameter R_m known as the "magnetic Reynolds number":

$$R_m = \frac{convective\ term}{diffusive\ term} = \mu_0\sigma UL$$

A system with a low magnetic Reynolds number will be dominated by scattering (usually collisional plasmas), while non-collisional plasmas will tend to be characterized by $R_m \gg 1$ and a dominance of convective effects on the evolution of the magnetic field. It should be noted that even for weakly resistive plasmas, the effect of field scattering becomes non-negligible when considering very small scales L; this is one of the reasons that makes magnetic reconnection possible at the "X points", discussed in section 4.6.2 of this chapter's Appendices.

We always keep the name "Ohm's law" for this equation, even in the non-collisional case $1/\sigma = 0$ (infinite conductivity), whereas it only includes the "change of reference" part. The consequences of Ohm's law are important in MHD (see section 4.6.2).

4.5.5. Conservation of energy

The last equation of the fluid part of the system is the kinetic energy transport equation. It shows that the kinetic energy is not conserved over time, the second member $\mathbf{j}.\mathbf{E}$ being a term of loss or gain. The reason for this non-conservation is the energy exchanges between field and charged particles, and it is verified that the total energy (kinetic + electromagnetic) corresponds to a conservative transport equation:

$$\partial_t \left(nm\frac{u^2}{2} + \frac{3}{2}p + \frac{B^2}{2\mu_0} \right) + \nabla. \left[\mathbf{u} \left(nm\frac{u^2}{2} + \frac{5}{2}p \right) + \frac{\mathbf{E}\times\mathbf{B}}{\mu_0} \right] = 0$$

In ideal MHD, the exchanges between kinetic energy and electromagnetic energy are strictly reversible. In the presence of a resistivity σ, a part of the transfers becomes irreversible (Joule effect).

4.5.6. MHD equilibrium conditions

The system of MHD equations we derived is adapted to the description of the phenomena slowly evolving over time. In particular, it makes it possible to describe the equilibrium situations of a magnetized plasma. This description is particularly important for the study of magnetic confinement fusion devices, the purpose of which is to constrain the dynamics of the plasma via the application of a magnetic field.

Given the equations derived in section 4.5.2, the equilibrium state, characterized by $\mathbf{u} = 0$ and $\partial_t = 0$, is described by the equations:

$$\mathbf{j} \times \mathbf{B} - \nabla p = 0$$

$$\nabla \times \mathbf{B} = \mu_0 \mathbf{j}$$

$$\nabla . \mathbf{B} = 0$$

$$\nabla . \mathbf{j} = 0$$

According to the first equation, the equilibrium state is found when the Laplace forces exerted by the magnetic field on the plasma are counterbalanced in every respect by the pressure forces internal to the plasma. Since the Laplace force has no component parallel to the field, it is easy to see that the pressure must be constant along the field lines: in an MHD equilibrium, the field lines are isobars. Using the notions of pressure and magnetic stress, we can rewrite the first two equations as:

$$F_T + \nabla \left(p + \frac{B^2}{2\mu_0} \right) = 0$$

illustrating another property: in the absence of curvature of the field lines (and therefore of magnetic stress force), the total pressure is constant at all points in an MHD equilibrium, which means that any variation of the kinetic pressure is compensated by an opposite variation of the magnetic pressure. In other words, a local pressure imbalance is necessarily compensated for by a stress force and thus by a curvature of the field lines.

4.5.7. *Examples of MHD balance*

Let us consider an infinite plasma column along the z axis. A magnetic field is applied in order to confine the column, that is, to prevent it from extending radially under the effect of its internal pressure p. Since the problem is invariant by rotation of angle θ and by translation along z, only the variable r will intervene in the description. The pressure force is thus radially exerted, and the confinement must be achieved via a magnetic field according to $\mathbf{e}_\theta$ and/or $\mathbf{e}_z$. We take $\mathbf{B}$ of the form $\mathbf{B} = B_\theta(r)\mathbf{e}_\theta + B_z(r)\mathbf{e}_z$, which imposes on the plasma a radial Laplace force equal to:

$$(\mathbf{j} \times \mathbf{B})_r = -\frac{\partial}{\partial r}\left(\frac{B_\theta^2}{2\mu_0} + \frac{B_z^2}{2\mu_0} \right) - \frac{B_\theta^2}{\mu_0 r}$$

where the decomposition of the Laplace force appears in terms of pressure and stress, as seen previously. Equilibrium of the column is thus obtained if the following equation between p and B is verified at all points:

$$\frac{\partial}{\partial r}\left(p + \frac{B_\theta^2}{2\mu_0} + \frac{B_z^2}{2\mu_0}\right) + \frac{B_\theta^2}{\mu_0 r} = 0$$

There are two classic ways to achieve such a balance.

The balance obtained by imposing a field along the column (along z), and therefore orthoradial currents, is called the "θ-pinch" configuration (the currents being along $\mathbf{e}_\theta$) (Figure 4.7). The equilibrium condition is then reduced to a simple pressure balance in every respect:

$$p(r) + \frac{B_z^2(r)}{2\mu_0} = \frac{B_{ext}^2}{2\mu_0}$$

where B_{ext} is the field outside the plasma column (where the pressure is zero).

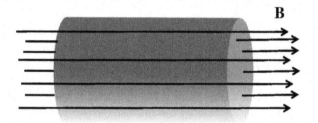

Figure 4.7. θ-pinch configuration; balance is assured by a magnetic field Bz (black) along the plasma column (purple). For a color version of this figure, see www.iste.co.uk/belmont/plasma.zip

The other possible configuration is that obtained by imposing an orthoradial field, via currents along z. This configuration is called "z-pinch", and its description is a little more complex, since it involves the magnetic stress (Figure 4.8). The equilibrium condition is then written as:

$$\frac{\partial}{\partial r}\left(p + \frac{B_\theta^2}{2\mu_0}\right) + \frac{B_\theta^2}{\mu_0 r} = 0$$

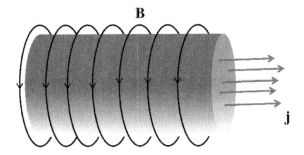

Figure 4.8. *Z-pinch configuration: the equilibrium is maintained by an orthoradial magnetic field B_θ (black), via currents along the plasma column (purple). For a color version of this figure, see www.iste.co.uk/belmont/plasma.zip*

It should be noted that, while it is possible to find conditions on the pressure and magnetic field profiles quite easily to obtain an equilibrium situation, it is not guaranteed that these equilibriums are stable. In this case, the two types of equilibrium described here are subject to instability.

Two examples of these instabilities for a z-pinch configuration are shown in Figure 4.9. A local tightening of the magnetic field lines (i.e. a local increase in the modulus of the magnetic field) will tend to amplify: this phenomenon is called sausage instability (the field lines strongly tighten at certain points, and the plasma column looks like a "bunch" of sausages). Another type of instability occurs when a displacement of the field lines with respect to the axis of symmetry takes place. The action of Laplace forces in this configuration tends to amplify the curvature of the plasma column. It is an instability called kink instability.

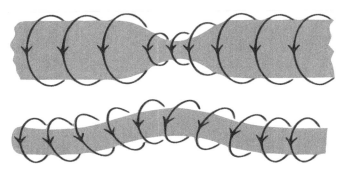

Figure 4.9. *Z-pinch configuration. Two examples of instability that will affect the equilibrium of a plasma column confined by an orthoradial field: (top) sausage instability and (bottom) kink instability*

The θ-pinch configuration is more stable compared to the field-line disturbances due to the stabilizing action of the magnetic stresses, shown in Figure 4.10: the light blue arrow indicates the restoring force of the field lines deformed by a disturbance. Other types of instability may, however, develop.

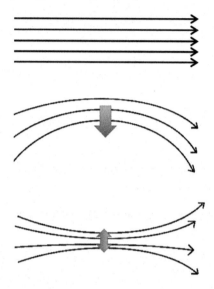

Figure 4.10. *Magnetic field for the θ-pinch configuration; the magnetic stress will compensate the disturbances of the field lines shown in the figure and bring them back to their original state (top). For a color version of this figure, see www.iste.co.uk/belmont/plasma.zip*

The control of these instabilities is an important issue for the development of magnetic plasma containment devices.

4.6. Appendices

4.6.1. *Transport equations in conservative form*

4.6.1.1. *Transport equations*

Many of the equations that have been manipulated in this chapter are of the form:

$$\partial_t(a) + \nabla.(\mathbf{b}) = c$$

Such equations are called transport equations.

If the volume density of a quantity A is represented by a, then the integration of this equation on a given volume indeed tells us that the temporal variation of A (integral of $\partial_t(a)$) is due, on the one hand, to a flow of this quantity through the outer boundaries of the volume (integral of $\nabla.(\mathbf{b})$ as "current" of the scalar quantity, $\mathbf{b}$ is homogeneous to $[a]$ * velocity) and, on the other hand, to a creation (or annihilation) of that quantity in the volume (integral of c). For example, we can think of the density of inhabitants in a country, which varies over time as a result of migratory flows across borders, and births and deaths in the country.

Putting all the equations of a system in this form, where possible, is extremely useful: it allows us to identify the "invariants" of the problem.

4.6.1.2. Conservative equations

A transport equation is said to be "conservative" if $c = 0$, that is, if there is no loss or gain in volume. This form of equation is used, for example, to find what is conserved at the crossing of a planar and stationary discontinuity like a shock wave. We show that an equation of the above form is satisfied (with $c = 0$) and results in the conservation $b_{n1} = b_{n2}$ of the normal component of b upstream and downstream of the discontinuity.

Most of the equations we wrote (in their so-called Eulerian form) were from the outset in an almost conservative form, with the exception of the momentum and energy equations for which a little extra work is needed.

4.6.2. Frozen field and reconnection

4.6.2.1. Ohm's law and frozen field

In ideal MHD, Ohm's law is simply written as:

$$\mathbf{E} = -\mathbf{u} \times \mathbf{B}$$

and it is used in the system only through the rotation of $\mathbf{E}$ (Faraday's law).

To this equation is associated the notion of magnetic reference frame (see Chapter 2), which confers an identity to each magnetic field line during the movement: Ohm's law shows that, in every point, the perpendicular velocity of a field line is equal to the fluid's velocity. It is said that the force line is "frozen" in the plasma, which means that it is driven and deformed in the movement of the fluid as a passive tracer.

4.6.2.2. *Conservation of connections*

The principle of the frozen field can be re-stated by saying that there is "conservation of connections", that is, in ideal MHD, two points initially connected by a field line (Figure 4.11) remain always connected during the movement. This means that a movement that transforms the figure on the left-hand side into that on the right-hand side is forbidden by the ideal MHD: we cannot "cut off" field lines.

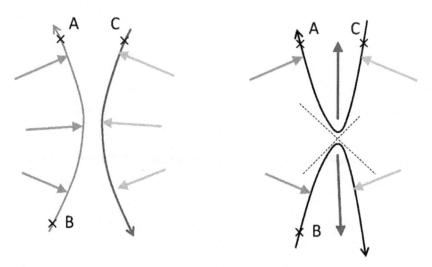

Figure 4.11. *Diagram of reconnection. (Left) A is connected to B but not to C. (Right) A is connected to C but not to B. In ideal MHD, no movement of plasma can lead to this change of connections. For a color version of this figure, see www.iste.co.uk/belmont/plasma.zip*

This theorem of the conservation of connections can be demonstrated by reasoning on an infinitesimal element of fluid having the form of a small segment oriented δl. Consider that this segment has its origin initially placed at a point O and that it is aligned with the magnetic field $\mathbf{B} = B\mathbf{b}$. In the course of time, the segment being driven by the flow, it moves and its direction changes. If we can show that it always remains aligned with the existing magnetic field at the point O, where its origin is, the extension to the whole of a field line will be obvious.

Let the vector $\delta \mathbf{C} = \delta \mathbf{l} \times \mathbf{B}$. Initially, we have $\delta \mathbf{C} = 0$, characteristic of an aligned segment. Let us calculate $d_t(\delta \mathbf{C})$ to verify that the segment remains aligned:

$$d_t(\delta \mathbf{C}) = d_t(\delta \mathbf{l}) \times \mathbf{B} + \delta \mathbf{l} \times d_t \mathbf{B} = \delta \mathbf{l}.\nabla(\mathbf{v}) \times \mathbf{B} + \delta \mathbf{l} \times [-\nabla \times \mathbf{E} + \mathbf{v}.\nabla \mathbf{B}]$$

By letting $E = -v \times B + E_{//}$:

$$d(\delta C) = B\delta l \left[\partial_{//}(\mathbf{v}) \times \mathbf{b} + \mathbf{b} \times (\partial_{//}(\mathbf{v}) - \nabla \times E_{//} / B) \right] = [\nabla \times \Delta E] \times \delta l$$

We verify that $d_t(\delta C) = 0$ in ideal MHD, since then $E_{//} = 0$; the law of conservation of connections is thus well established.

When non-ideal terms, for example, resistivity, are taken into account, this adds a non-null $\mathbf{E}_{//}$ term to Ohm's law. The previous calculation of $d_t(\delta C)$ then makes it possible to quantify the "reconnected" flow by unit of time.

4.6.2.3. Conservation of flux

A force tube is characterized by a certain value of the magnetic flux. It can be shown that, in ideal MHD, this value remains constant during the movement of the flux tube. This is obviously related to the conservation of flux through the cyclotron trajectory of individual particles (see Chapter 2).

4.6.2.4. Reconnection

A frequently studied case is one where the ideal MHD is valid almost everywhere except for a small region of space, typically an "X-point" like that shown in Figure 4.11[8].

The problem of reconnection is important for the following reason: magnetic configuration, where all field lines are anti-parallel, like those in Figure 4.11 (left), has a magnetic energy greater than the right configuration, where some lines are connected. If the ideal MHD remains valid everywhere, then it prohibits the system, for topological reasons, from evolving towards its minimum energy configuration. The lines pile up in a vertical layer in the center of the figure, and we thus reach a stationary state, which can be described as "meta-stable". If, on the contrary, the

---

8 The breaking of ideality can, for example, be due to a value of B which vanishes at the point in X: this has the effect of canceling the term of order 0 in the development in slow variations (it is understandable that MHD cannot be valid in the absence of a magnetic field); it may also be due to other local conditions, with B not equal to zero, but this is beyond the scope of this book and we will not show it here. Be that as it may, the velocity gradients are very strong around the X-point, which makes the usually negligible terms all the more important: the terms of viscosity or resistivity if the medium is weakly collisional, all of the terms neglected because of "slow variations" for collisionless plasmas.

constraint of ideality "breaks" at one point, then there is reconnection at this point and we witness a rapid transition to the minimum energy state: magnetic energy finally has the topological possibility of diminishing. During the transition, it is essentially converted into directed kinetic energy, via the magnetic stress forces (vertical flows in Figure 4.11), a part that can also be dissipated as thermal energy[9].

9 This phenomenon is sometimes invoked to explain the heating of the solar corona.

5

Waves in Plasmas in
the Fluid Approximation

In all areas of physics (mechanics, electronics, etc.), we know that many properties of a system are determined by its linear *behavior*. When this system is governed by an ordinary differential equation dealing with time, this, first of all, involves determining its *own modes of oscillation* (solid body subjected to a restoring force, RLC oscillator, etc.). When the system is governed by a differential equation with partial derivatives on time and space, this amounts to looking for its *own modes of propagation* (sound waves, evolution of potential on a power line, etc.). These results provide the first and indispensable information on how the system reacts: wave dispersion relation, stability of the equilibrium, etc. They will also allow us to introduce the notions of cutoff frequency and resonance frequency, and to give an example of application of the so-called "frozen-in" theorem, that is, the fact that the magnetic field lines are frozen into the fluid plasma. We will apply the classical method of computation of the propagation modes on some examples of waves and we will present the case of an instability.

5.1. Waves in plasmas: modes of propagation

In plasma as in any other medium, any fluctuation of small amplitude propagates according to linear modes (eigenmodes) of propagation. In a neutral gas, for example, if we initially create a small plane layer of over-density, that is, a small bump on the curve n(x), it will break down into two smaller bumps, each one propagating at the speed of sound, one in one direction and the other in the opposite direction (Figure 5.1). These two small bumps will be of the same shape and the same amplitude (and therefore half of the initial bump) if we do not introduce an initial velocity perturbation. This is consistent with the eigenmodes calculations of the linear wave equation, as we will see later, since there are two eigenmodes of propagation which are the two sound waves, one in each direction.

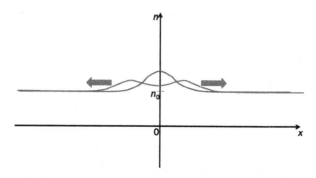

Figure 5.1. *Density perturbation in a neutral gas. The initial perturbation (in blue) breaks down into two perturbations propagating in opposite directions at the speed of sound (in red). For a color version of this figure, see www.iste.co.uk/belmont/plasma.zip*

The method presented in this chapter actually applies to the propagation of a sinusoidal wave and not of a random perturbation like this bump of localized density. However, this is sufficient because we know how to decompose an initial perturbation in a sum of sinusoids, as in a wave packet. This is the principle of the Fourier transform[1]. In this example of sound waves, all the components going in the same direction propagate at the same phase velocity c_s (speed of sound) or $-c_s$. Once the packet going to the right is separated from the packet going to the left, they both propagate at speed c_s, without being deformed. The same thing will happen with all the so-called "non-dispersive" modes, that is, such that the angular frequency of the wave ω and its wave vector $-k$ are linked by an equation of the form $\omega/k = $ cst.

The eigenmodes of a system are monochromatic solutions characterized by sinusoidal variations in space (k) and time (ω). There is usually a finite number of solutions because we find that the set of equations can be verified simultaneously for a given k only if ω respects an equation that links the wave vector and the frequency, known as a dispersion equation $D(\omega,k) = 0$. The same method for determining eigenmodes can be applied to plasma. In this chapter, we will assume that the considered plasma can be correctly described by fluid equations as introduced in Chapter 4. Each fluid system of this type has different solutions, that is, different modes, the dispersion and polarization of which can be determined. In a

---

1 Since the problem is linear (small perturbations), each Fourier component propagates independently of the others and the same linear superposition gives the propagation of the set (inverse Fourier transformation).

plasma, there are several solutions to the fluid system that is solved, and the number of linear eigenmodes is generally greater than that in the case of a gas. The number of modes is always even because the two directions of propagation are always possible. Which modes will be excited among all these modes depends on a larger number of initial conditions than for the gas: it depends on the initial disturbances of density and velocity, but also on the disturbances of the charge density, magnetic field, etc.

Before explicitly searching for eigenmodes, it is useful to try to figure out how many they are. Can we predict the number of possible solutions in ω for a sinusoidally varying solution in space, monochromatic in k? The answer is yes: the number of eigenmodes is equal to the order of the differential system in time that we solve. In other words, for a differential system consisting of N first-order equations in time and interconnecting N variables, the number of solutions to the problem posed is N.

This can easily be verified for the environments that we know:

– one-dimensional neutral medium: 2 variables n and u → 2 modes (sound waves in both directions);

– 1D electrostatic plasma with immobile ions: 2 variables n_e and u_e→ 2 modes (Langmuir waves or plasma oscillations in both directions);

– Plasma in 3D MHD: 6 variables n, $\mathbf{u}$, $\mathbf{B_t}$→ 6 low-frequency modes;

– 3D bi-fluid plasma: 12 variables n_1, n_2, $\mathbf{u}_1$, $\mathbf{u}_2$, $\mathbf{B_t}$, $\mathbf{E_t}$→ 12 modes: 6 at high frequency and 6 at low frequency.

It will be noted that the low-frequency modes in magnetized plasma can be calculated more simply from the MHD limit ("mono-fluid") presented in chapter 4.

NB. In this enumeration, **B** and **E** count only for two variables because only their components that are perpendicular to **k** are derived with respect to time: their longitudinal components are determined by $\nabla.\mathbf{B} = 0$ and $\nabla.\mathbf{E} = \rho / \varepsilon_0$, which are not differential equations in time. Similarly, the pressures p_e and p_i do not count in the enumeration, if their closure equations are not time dependent (e.g. $p = p(n)$).

5.2. Calculation of the propagation eigenmodes: classical method

The technique, to whatever system it is applied, has four stages. It has already been used in Chapter 1 for the very simple case of plasma oscillation.

5.2.1. *Writing the system of differential equations*

The medium to be analyzed is characterized by a system of differential equations with respect to time and space. It is without source terms on the right-hand side, that is, without "forcing" or excitation of the eigenmodes. A well-known example of general physics is the system of equations governing the propagation of pressure perturbations in air (sound waves). In plasmas, we will see that there are indeed a greater number of wave solutions because of the greater number of variables in the system, and that the number depends on the detail and the approximations considered in the description. The medium is also characterized by its equilibrium state: magnetized or non-magnetized plasma, at rest or in motion, cold or hot.

5.2.2. *Linearization*

The previous system is linearized in the vicinity of a "zeroth order" solution, or equilibrium state. We decompose all the variables n as the sum of their value at zeroth order, plus an extra term:

$$n = n_0 + n_1$$

with the assumption that $n_1 \ll n_0$. The variables n_0 are assumed to be known, and they satisfy the system of equations. If our system is described by a multivariate equation, that we write, for example, formally $E\ (n,\ \boldsymbol{u},\ p) = 0$, the equilibrium satisfies $E\ (n_0,\ u_0,\ p_0) = 0$: all the terms of the expansion at zeroth order satisfy exactly this equation. The linearization of the system then involves retaining only the first order of this expansion (power 1). The result is a linear equation in n_1, $\boldsymbol{u}_1$, p_1.

For example:

$$\partial_t(nm\mathbf{u}) + \nabla.(nm\mathbf{uu} + \mathbf{p}) \xrightarrow{\textit{linearization}} \partial_t(n_1 m\mathbf{u}_o + n_o m\mathbf{u}_1)$$
$$+\nabla.(n_1 m\mathbf{u}_o \mathbf{u}_o + n_o m\mathbf{u}_1 \mathbf{u}_o + n_o m\mathbf{u}_o \mathbf{u}_1 + \mathbf{p}_1)$$

5.2.3. *Research of complex exponential solutions ("algebraization")*

We look for solutions of the system, which are of the type "progressive and monochromatic planar waves", that is, solutions for which the variations of all the variables of the system are of the form:

$$e^{-i\omega t + i\mathbf{K}\cdot\mathbf{r}}$$

An algebraic system is thus obtained instead of the differential system. Simply replace:

$$\partial_t \rightarrow -i\omega \text{ and } \nabla \rightarrow i\mathbf{k}$$

When we have determined the particular solutions that verify this algebraic system, we will admit that the linearized differential system is solved because the general solution is always a linear combination of these particular solutions (the coefficients being determined by the boundary conditions of the problem). It is therefore assumed that exponential solutions form a complete subset and a basis for all solutions. The proof of this statement and its limits of validity can be studied only by making use of the Fourier and Laplace transformations; we will not do it in this chapter because it is superfluous for the moment. However, we will come back to this point in Chapter 6 (study of kinetic processes) since these questions are essential for obtaining and understanding the kinetic results.

5.2.4. Resolution

The algebraic system above is a homogeneous system (i.e. without terms on the right-hand side). It can be expressed in the form of a matrix applied to the variables of the system. In general, for any ω and $\mathbf{k}$, it admits only the trivial solution for which all the linear variables are zero (no wave). To obtain solutions other than the trivial solution, the variables ω and $\mathbf{k}$ need to verify an equation $\boxed{D(\omega, \mathbf{k}) = 0}$ which ensures the cancellation of the determinant of the matrix of the algebraic system. This equation can also be obtained by solving the system by replacing the variables and simplifying the equations, without explicitly calculating the determinant. We thus obtain the *dispersion equation* of the medium. Each solution of the dispersion equation is associated with a direction in the space of the variables (or with a subspace of dimension> 1 in case of degeneracy), just as, for a matrix, each eigenvalue is generally associated with a set of collinear eigenvectors. The eigenvectors define the *polarizations* (in the broad sense) of the different solutions.

If the physical system contains sources or forcing terms, the result of the calculation above still provides important information: it is known that the forcing effect will be at maximum when these source terms follow the dispersion equation given above.

In the following sections, we will familiarize ourselves with this method through several examples.

5.3. Fluid treatment of the plasma wave

The plasma oscillation has already been introduced in Chapter 1 in the case of a zero-temperature plasma; we will now include the effects of temperature. This type of wave is also called "Langmuir wave" or "high-frequency wave" or "electronic wave" depending on the context. Indeed, as already discussed, in plasma, the two main species that appear, ions and electrons, have a very different mass. If the ions are protons, the mass ratio is given by $m_e/m_p = 1/1836$[2]. For this reason, if we consider fast oscillations, the electrons that are the lightest will be set in motion, but not the ions that behave like a neutralizing and motionless background. It is therefore easy to generalize the calculation of the plasma oscillation in the context of waves in the fluid description of plasma. This gives the first application of the technique presented in the previous section.

If we consider a hot plasma, not magnetized, and at rest (null average speed), its equilibrium is given by $Zn_{i0} = n_{e0}$, $\mathbf{u}_{e0} = \mathbf{u}_{i0} = 0$, $T_e = T_{e0}$, $T_i = 0$, $\mathbf{E}_0 = 0$, $\mathbf{B}_0 = 0$.

As in the study of plasma oscillations in Chapter 1, we make the hypothesis that the perturbations studied are associated with a unidirectional movement that we choose as x, and all the variables depend in space only on x and are oriented along which we define the x axis. In addition, we make the assumption that, for the case of hot plasma, similar to what was found in Chapter 1, the magnetic field perturbation is zero and the wave consists of purely electrostatic perturbations. In fact, for longitudinal waves (propagation of the wave parallel to the direction of the field), the magnetic field is always zero. On the other hand, the dispersion equation will be different from that of the cold plasma; in particular, we will see that, in the hot plasma, the frequency depends on the wave vector.

We write $n_e = n_{e0} + n_{e1}$, $u_{ex} = u_{e1x}$, $E = E_{1x}$, $T_e = T_{e0} + T_{e1}$ and remember that $p_e = n_e k_B T_e = p_{e0} + p_{e1} = n_{e0} k_B T_{e0} + n_{e0} k_B T_{e1} + n_{e1} k_B T_{e0}$. Because of the chosen plasma conditions, all the quantities with index 0 are constants, and the quantities with index 1 depend on x and t.

---

2 The parameter that comes into play to be more precise is Zm_e/Am_p, where Z is the ionic charge and A is the atomic mass.

There is no perturbation for ions that are considered a still background; as an exercise, the reader can show that the modification induced to the dispersion equation because of the ions' movement is negligible. The first two fluid equations of the system for the electrons are (see chapter 4):

$$\partial_t(n_e) + \nabla \cdot (n_e \mathbf{u}_e) = 0$$
$$m_e \partial_t (n_e \mathbf{u}_e) + m_e \nabla \cdot (n_e \mathbf{u}_e \mathbf{u}_e) = -\nabla(p_e) - en_e(\mathbf{E} + \mathbf{u}_e \times \mathbf{B})$$

If we write the variables explicitly, we obtain:

$$\partial_t(n_{e0} + n_{e1}) + \nabla \cdot ((n_{e0} + n_{e1})\mathbf{u}_{e1}) = 0$$
$$m_e \partial_t ((n_{e0} + n_{e1})\mathbf{u}_{e1}) + m_e \nabla \cdot ((n_{e0} + n_{e1})\mathbf{u}_{e1}\mathbf{u}_{e1}) = -\nabla(p_{e1}) - e(n_{e0} + n_{e1})\mathbf{E}_1$$

where it is recalled that the component of the zeroth order velocity is zero (the plasma is at rest).

The derivatives of the quantities with index 0 are zero (since we have constant density and pressure at equilibrium), and the quadratic terms in the quantities of order 1 are very small with respect to the same quantities to the first power. We can therefore neglect them in these equations, and we finally obtain the linearized equations (see section 5.2.2) for the given geometry:

Number of particles conservation: $\partial_t(n_{e1}) + n_0 \partial_x(u_{e1x}) = 0$ [5.1]

Momentum conservation: $n_0 m_e \partial_t(u_{e1x}) + \partial_x(p_{e1}) = -n_0 e E_{1x}$ [5.2]

To "close" the system, it is necessary to express the pressure as a function of density and/or velocity. Finding a correct "closure equation" for the system is a subtlety of the fluid method used and the justification for such an equation, if it exists, cannot be found without returning to the kinetic equations. The following intuitive argument can be used: if the motion is fast, the compressed electrons do not have time to exchange energy with their neighbor in the dilated part, and as in an adiabatic disturbance in a gas, the temperature increases where there is compression, and decreases if there is dilation. The adiabatic closure equation (introduced in Chapter 4) can then be used and the linearized version of this equation is:

$$p_{e1}/p_{e0} = \gamma n_{e1}/n_{e0} \Leftrightarrow p_{e1} = \gamma k_B T_{e0} n_{e1} = \gamma m_e V_{the}^2 n_{e1}$$ [5.3]

and the thermal velocity is introduced $V_{the}^2 = k_B T_{e0}/m_e$. This is the simplest closure equation (except for the cold plasma assumption $p_e = p_{e0} = 0$). The complete kinetic calculation will show a posteriori that this hypothesis is justified as long as the wavelengths are sufficiently large (relative to the Debye length) and that, in this case, $\gamma = 3$, which corresponds to an adiabatic movement along one dimension (remember that for a perfect gas, if one has d degrees of freedom, $\gamma = (d+2)/d$).

If we take the partial derivative with respect to t in equation [5.1] and the partial derivative with respect to x in equation [5.2], after replacing p_{e1} as a function of n_{e1} by equation [5.3], we obtain two equations for n_{e1} and u_{ex1} where we can easily eliminate u_{ex1}. We thus obtain the new equation governing the variations of densities due to the electric field:

$$(\partial_t^2 - 3 V_{the}^2 \partial_x^2)(n_{e1}) = \frac{n_0 e}{m_e} \partial_x (E_{x1}) \tag{5.4}$$

One can eliminate the electric field thanks to the Gauss–Maxwell equation, $\nabla . \mathbf{E} = e(Zn_i - n_e)/\varepsilon_0$. In the zeroth order, this equation corresponds to $\mathbf{E}_0 = 0$; $n_{e0} = Zn_{i0}$, and once linearized it takes the form:

$$\partial_x (E_{x1}) = -en_{e1} / \varepsilon_0 \tag{5.5}$$

The equation governing the evolution of density is now written:

$$\boxed{(\partial_t^2 + \omega_{pe}^2 - 3 V_{the}^2 \partial_x^2)(n_{e1}) = 0} \tag{5.6}$$

In addition to the terms already found in the cold plasma case in Chapter 1, we see that this time we have a term V_{the}^2 that couples the temporal oscillation with the spatial variation. This changes the characteristics of the plasma waves.

The wave dispersion equation is obtained by looking for solutions in the form of monochromatic plane waves $e^{i(kx - \omega t)}$ (see section 5.2.3), and equation [5.6] becomes:

$$(-\omega^2 + \omega_{pe}^2 + 3k^2 V_{the}^2)(n_{e1}) = 0$$

The resolution (see section 5.2.4) is in this case particularly simple: we have a non-trivial solution ($n_{e1} \neq 0$) only if ω and k are linked by the dispersion equation (see figure 5.2):

$$\omega^2 = \omega_{pe}^2 + 3k^2 V_{the}^2$$

which is the dispersion equation of plasma waves or Langmuir waves. Thus, we see all the characteristic parameters of the phenomenon: angular frequency ω_{pe}, limit phase velocity for large k $\sqrt{3}V_{the}$, and characteristic length in the vicinity of which the temperature effects become evident $\lambda_d = V_{the}/\omega_{pe}$ (Debye length). We also see that there is energy propagation, since the group velocity is no longer zero. In particular, group velocity v_g and phase velocity v_φ are inversely proportional, according to $v_g v_\varphi = 3V_{the}^2$. In the limit where the temperature of the plasma is low $3k^2 V_{the}^2 \ll \omega_{pe}^2$, we find as expected the result of Chapter 1 and the dispersion equation for a cold plasma, $\omega = \omega_{pe}$.

This simple fluid result is correct for $k \ll 1/\lambda_d$ (fairly large scales or plasma that is not too hot); it becomes less correct for larger values of k: it can only be considered as the first order of an expansion in $k\lambda_d$. When k approaches $1/\lambda_d$, not only is the value of ω different from the one calculated above, but the very fact of calculating monochromatic plane waves in $e^{-i\omega t}$ becomes impossible. The complete kinetic calculation, in chapter 6, will show that for these wavelengths, in a plasma with an initial Maxwellian velocity distribution, the only solutions that exist are damped waves (by "Landau damping"), that is, solutions in $e^{-i\omega t - \eta}$. No fluid theory can lead to such a result and we shall see later that the calculation which shows this gives a particularly important role to the so-called "resonant" particles, that is, those whose velocity is equal to the phase velocity.

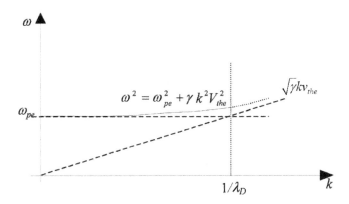

Figure 5.2. *Dispersion equation for hot plasma (fluid theory)*

5.4. Example of "electron" instability: two-stream instability

The systems of fluid equations we have just studied, as well as others that we will see later, have real eigenmodes (real ω for real $\mathbf{k}$), which means that they give rise to constant amplitude waves when the medium is linearly perturbed. These waves will actually propagate without damping or amplifying until they are eventually damped by effects not taken into account in the system (collisions or Landau effect already mentioned). But there are other cases where some of these waves can be unstable. This is particularly the case if there is a non-zero relative velocity between the different populations present in the plasma. This means that the perturbation grows and loses its waveform quickly, and as a consequence the framework of linear theory is not valid anymore and the system evolves until it reaches a new nonlinear equilibrium. There are several examples of instability of this type. The simplest that will be detailed here is what we call the two stream instability, which is an electrostatic instability (no perturbation of the magnetic field) and "electronic" (no movement of the ions, which stay as a neutralizing still background). It can develop in cold plasma (where we can neglect the effects of temperature). The system is composed of two identical (a and b) and cold electron beams which propagate in opposite directions with the same absolute velocity in a background of neutralizing ions (this system has already been introduced in Chapter 4, Figure 4.1). The equilibrium of the system is given by:

$$Zn_{i0} = n_{e0}^a + n_{e0}^b = n_{e0}, n_{e0}^a = n_{e0}^b = \frac{n_{e0}}{2}, \mathbf{u}_{e0}^a = \mathbf{v}_0, \mathbf{u}_{e0}^b = -\mathbf{v}_0, \mathbf{u}_i = 0,$$

$$T_e^{a,b} = 0, T_i = 0, \mathbf{E}_0 = 0, \mathbf{B}_0 = 0$$

The system to be analyzed is:

$$\partial_t n_e^a + \mathbf{\nabla} \cdot (n_e^a \mathbf{u}_e^a) = 0 \qquad [5.7]$$

$$m_e \partial_t (\mathbf{u}_e^a) + \mathbf{u}_e^a \cdot \mathbf{\nabla}(\mathbf{u}_e^a) = -e\mathbf{E} \qquad [5.8]$$

$$\partial_t n_e^b + \mathbf{\nabla} \cdot (n_e^b \mathbf{u}_e^b) = 0 \qquad [5.9]$$

$$m_e \partial_t (\mathbf{u}_e^b) + \mathbf{u}_e^b \cdot \mathbf{\nabla}(\mathbf{u}_e^b) = -e\mathbf{E} \qquad [5.10]$$

$$\mathbf{\nabla} \cdot \mathbf{E} = e(Zn_i - n_e^a - n_e^b) \qquad [5.11]$$

for a purely electrostatic perturbation and for a cold plasma. As in the previous sections, we consider small perturbations, $n_e^a = n_{e0}^a + n_{e1}^a$, $n_e^b = n_{e0}^b + n_{e1}^b$ with $n_{e1}^a \ll n_{e0}^a$, $n_{e1}^b \ll n_{e0}^b$, etc.,; the system is linearized and monochromatic plane wave solutions are sought. We will assume that the propagation direction of the beams is along the x axis. In this case, the planar solutions are of the form $e^{i(kx - \omega t)}$

for all the variables. The electrostatic solutions are longitudinal, so the electric field and the velocity perturbation are parallel to the direction of propagation and parallel to each other. There is only the x component of the field. The system then becomes an algebraic system:

$$-i(\omega - kv_0)n_{e1}^a + ik\frac{n_{e0}}{2}v_{e1}^a = 0$$

$$-im_e(\omega - kv_0)v_{e1}^a = -eE_1$$

$$-i(\omega - kv_0)n_{e1}^b + ik\frac{n_{e0}}{2}v_{e1}^b = 0$$

$$-im_e(\omega + kv_0)v_{e1}^b = -eE_1$$

$$-ikE_1 = -e(n_{e1}^a + n_{e1}^b)$$

If we combine these equations we find the dispersion equation:

$$1 - \frac{\bar{\omega}_{pe}^2}{(\omega - kv_0)^2} - \frac{\bar{\omega}_{pe}^2}{(\omega + kv_0)^2} = 0 \qquad [5.12]$$

where the plasma frequency associated with a single beam has been introduced:

$$\bar{\omega}_{pe}^2 = \frac{e^2 n_{e0}}{2m_e\varepsilon_0}$$

We note that in the limit where v_0 tends to 0, equation [5.12] corresponds to the plasma oscillations and the dispersion equation is $\omega^2 = 2\bar{\omega}_{pe}^2 = \omega_{pe}^2$.

If we develop the dispersion equation, we obtain a quadratic equation in ω^2 that we can solve simply. If $kv_0/\bar{\omega}_{pe} > \sqrt{2}$, the solutions of this equation are four real roots $\omega_{1,2} = \pm\omega_a$, $\omega_{3,4} = \pm\omega_b$, and they correspond to two oscillatory modes with characteristic frequencies ω_a and ω_b. It is useful to introduce normalized variables Ω $\omega/\bar{\omega}_{pe}$ and $K = kv_0/\bar{\omega}_{pe}$: the two solutions for the characteristic frequencies are given by $\Omega_{a,b} = \sqrt{1 + K^2 \pm \sqrt{1 + 4K^2}}$. The solution for density (or electric field or velocity) perturbation, in this case, is the sum of the two oscillatory modes:

$$n_{e1} = \tilde{n}_{e1a}e^{i(kx - \omega_a t)} + \tilde{n}_{e1b}e^{i(kx - \omega_b t)} + c.c.$$

where c.c. indicates the conjugate complex.

If, on the other hand, $kv_0/\bar{\omega}_{pe} < \sqrt{2}$, there are two purely real solutions $\omega_{1,2} = \pm\omega_a$ and two purely imaginary solutions $\omega_{3,4} = \pm i\gamma$ with $\Omega_b = i\Gamma =$

$i\sqrt{1+K^2} - \sqrt{1+4K^2}$. It is therefore possible to have an instability, that is, a mode whose amplitude increases with time. In this case, the solution for the density perturbation (or electric field or velocity) takes the form:

$$n_{e1} = \tilde{n}_{e1a}e^{i(kx-\omega_a t)} + \tilde{n}_{e1c}e^{ikx+\gamma t} + \tilde{n}_{e1d}e^{ikx-\gamma t} + \tilde{n}_{e1a}^*e^{-i(kx-\omega_a t)}$$

and very quickly the term proportional to $e^{\gamma t}$ will be dominant over the others. After a time of about $1/\gamma$, the perturbation will become large and comparable to the equilibrium density n_{e0}: the linear theory then no longer applies.

To know how the evolution evolves, we can solve the nonlinear system numerically. This is shown in Figure 5.3, in which three images are reproduced at three different times of the electric field, the density perturbation normalized to the equilibrium density n_{e1} / n_{e0}, and the velocity (normalized to the speed of light) of the two beams is positive for one beam and negative for the other. They have been obtained by the numerical simulation of the two stream instability (hence, in the case $kv_0/\overline{\omega}_{pe} < \sqrt{2}$). In the simulation, the equations are not linearized, and the self-consistent evolution of the beam/electric field system is solved numerically. The initial perturbation is very small, and for some time it remains weak; the simulation therefore reproduces the results of the linear theory. Over longer times, the nonlinear effects and the evolution of the velocity distribution of the particles are taken into account by the simulation, and allow us to understand what happens when the linear theory fails.

It can be seen that, at the beginning (figure 5.3a), the density and field perturbations have a sinusoidal shape and are very small, of the order of 10^{-3}. The velocity perturbation, which is also very small, cannot be distinguished in the figure. After a certain time (Figure 5.3b), density, field, and velocity perturbations have grown considerably due to the exponential term, and lost their sinusoidal shape: the nonlinear terms that have been neglected in the solution of the linearized equations have become important. The modification of the initial velocities then becomes visible. The disturbance is so great that, at a certain moment (figure 5.3c), it causes a mixing of the beams and an enlargement of the possible velocities of the particles, which evokes a notion of temperature. Once the beams are mixed, the system becomes stable again. The original equilibrium has been so modified that the analysis that has been done no longer applies and the system has found a new equilibrium. In this regard, it is important to remember that in order to have the possibility of developing an instability, the system must have so-called free energy. The growing field and density perturbations gain a certain amount of energy: it is the initial kinetic energy of the beams, which is converted into energy of the instability; once the beams no longer exist, the instability saturates and the growth stops.

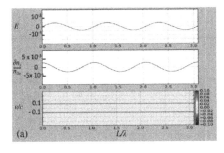
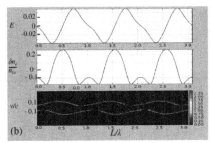
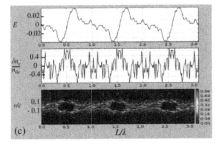

Figure 5.3. *Electric field, density perturbation and velocity of the two electron fluids as a function of the space at three different times, which illustrate the growth and saturation of the double beam instability. For the velocity, the color scale is proportional to the fraction of particles that have a given velocity. For a color version of this figure, see www.iste.co.uk/belmont/plasma.zip*

5.5. Other "electronic" propagation modes

The first simplification in the calculation of the preceding section consisted in neglecting the ion movement. This is justified *a posteriori* by the result obtained since, at frequencies of the order of (or greater than) the plasma frequency ω_{pe}, the ions, which are too heavy, can actually be considered as immobile (in particular $\omega_{pe} \gg \omega_{pi}$). We can study other high-frequency modes, which are also called "electronic" modes. In this section, we will investigate whether there are non-electrostatic solutions in this high-frequency range. If the plasma has a very low density and $\omega \gg \omega_{pe}$, we can guess, for example, that we will find the usual electromagnetic waves in "vacuum" for which $\omega = kc$. This is however only an approximate solution and we can calculate the general dispersion relation of electromagnetic waves in plasma. Since ions are considered immobile, as already seen, we do not need to write equations for the evolution of ions, and the only equation involving ions is the condition of global neutrality at equilibrium, $Z n_{i0} = n_{e0} = n_0$, which we will use in Maxwell's equations if necessary.

5.5.1. *System of equations*

The system of equations to be analyzed for an electromagnetic solution is:

$$\begin{cases} \partial_t(n_e) + \nabla \cdot (n_e \mathbf{u}_e) = 0 \\ n_e m_e \partial_t(\mathbf{u}_e) + n_e m_e \mathbf{u}_e \cdot \nabla(\mathbf{u}_e) + \nabla(p_e) = -n_e e(\mathbf{E} + \mathbf{u}_e \times \mathbf{B}) \\ d_t(p_e/n_e^{\gamma}) = 0 \end{cases}$$

$$\begin{cases} \nabla \cdot \mathbf{E} = -e\dfrac{n_e - Zn_i}{\varepsilon_0} \\ \nabla \times (\mathbf{E}) = -\partial_t(\mathbf{B}) \\ \nabla \times (\mathbf{B}) = -\mu_0 e(n_e \mathbf{u}_e - Zn_i \mathbf{u}_i) + \dfrac{1}{c^2}\partial_t(\mathbf{E}) \end{cases}$$

The fluid equations used are identical to those studied in the previous section, with the exception of the term in $\mathbf{u}_e \times \mathbf{B}$ introduced in the momentum conservation equation that must be kept in principle because now we are looking for solutions with magnetic field perturbations. Consequently, since the variable $\mathbf{B}$ is present in the system, it is necessary to use the set of Maxwell's equations in order to complete the system and not just the Maxwell–Gauss[3] equation. Note that for these frequencies, the displacement current cannot be neglected. Note also that a scalar pressure and a polytropic closure equation have been assumed. As has been said, these approximations will prove correct *a posteriori* (with $\gamma = 3$) as long as $k \ll 1/\lambda_D$.

5.5.2. *Electromagnetic waves in non-magnetized cold plasma: cutoff frequency*

We consider cold plasma, that is, where we can neglect the effects of temperature. This amounts to postulating $Te = 0$ everywhere. Moreover, the equilibrium considered is the same as in the previous section: $Zn_{i0} = n_{e0}$, $\mathbf{u}_{e0} = \mathbf{u}_{i0} = 0$, $T_i = 0$, $\mathbf{E}_0 = 0$, $\mathbf{B}_0 = 0$.

---

3 In practice, the equation relating to the divergence of the electric field is redundant. For an electrostatic electronic mode, we can use equivalently the Maxwell–Gauss equation or the Maxwell–Ampere equation. For an electromagnetic mode, we must consider the Maxwell–Ampere equation and the equations relating to the divergences are valid as long as they are verified in the initial condition.

In this case, we have only two equations for the description of the plasma, which will couple density and velocity of the electronic fluid, and the linearized equations for the variables n_{e1} and $\mathbf{u}_{e1}$ (see section 5.2.2) are given below. As these are electromagnetic modes, we must introduce perturbations of the electric and magnetic field: we will also have the variables $\mathbf{E}_1$ and $\mathbf{B}_1$. The term $(\mathbf{u}_{e1} \times \mathbf{B}_1)$ is neglected because it is of order two.

$$\partial_t (n_{e1}) + n_{e0} \nabla.(\mathbf{u}_{e1}) = 0 \qquad\qquad [5.13]$$

$$m_e \partial_t (\mathbf{u}_{e1}) = -e(\mathbf{E}_1) \qquad\qquad [5.14]$$

$$\nabla \times (\mathbf{E}_1) = -\partial_t (\mathbf{B}_1) \qquad\qquad [5.15]$$

$$\nabla \times (\mathbf{B}_1) = -\mu_o en_{e0} \mathbf{u}_{e1} + \frac{1}{c^2} \partial_t (\mathbf{E}_1) \qquad\qquad [5.16]$$

Maxwell's equations [5.15] and [5.16] require these fields to be orthogonal to each other and orthogonal to the direction of propagation, since the velocity $\mathbf{u}_{e1}$ in [5.16] is parallel to the electric field by equation [5.14]. We can deduce in particular that div $(\mathbf{E}_1) = 0$, and therefore, that there is no density perturbation associated with this mode.

We can combine the last two Maxwell equations to obtain a wave propagation equation by replacing the variable $\mathbf{B}_1$:

$$\partial^2_t (\mathbf{E}_1) - c^2 \Delta(\mathbf{E}_1) = \frac{en_{e0}}{\varepsilon_0} \partial_t \mathbf{u}_{e1} \qquad\qquad [5.17]$$

We therefore have three equations for three unknowns, n_{e1}, $\mathbf{u}_{e1}$, $\mathbf{E}_1$, and we obtain the wave dispersion equation by looking for a monochromatic plane wave solution of the form $e^{i(kx-\omega t)}$ for all the variables (see section 5.2.3). The system of equations [5.13], [5.14] and [5.16] becomes a homogeneous algebraic system:

$$-i\omega n_{e1} + in_{e0}\mathbf{k}.\mathbf{u}_{e1} = 0$$

$$-i\omega m_e \mathbf{u}_{e1} = -e\mathbf{E}_1$$

$$-\omega^2 \mathbf{E}_1 + c^2 k^2 \mathbf{E}_1 = -i\omega \frac{en_{e0}}{\varepsilon_0} \mathbf{u}_{e1}$$

If we examine the first equation, we find that the velocity perturbation being parallel to the electric field and orthogonal to the wave vector, the dot product (second term of the equation) is zero. As a result, the density perturbation is zero: there is no density perturbation for an electromagnetic wave that propagates in non-magnetized plasma. As anticipated, the charge density $-en_1$ is zero, which corresponds to a transverse electric field with div $(\mathbf{E}_1) = 0$ that is $\mathbf{k}.\mathbf{E}_1 = 0$.

The resolution of the system of the two remaining equations can be done by replacement (see section 5.2.4): we calculate $\mathbf{u}_{e1}$ starting from the second equation and replace it in the third one. We obtain the dispersion relation:

$$\boxed{\omega^2 = k^2c^2 + \omega_{pe}^2} \tag{5.18}$$

The dispersion relation could have been obtained equivalently by writing the homogeneous system in the form of a matrix and imposing the determinant as zero.

It is clear from the dispersion equation that the frequencies of the electromagnetic waves in the plasma are always greater than ω_{pe}, so the high frequency hypothesis that we made at the beginning that allowed us to treat the ions as immobile is well justified.

For $\omega \gg \omega_{pe}$, we find the well-known dispersion $\omega = kc$ of electromagnetic waves in vacuum. For high frequencies, plasma behaves as a very rarefied medium close to vacuum. For decreasing frequencies, we move away from this vacuum solution until the limiting value $\omega = \omega_{pe}$ is reached. By analogy with the dielectrics, the optical index of the medium can be introduced as the ratio between the phase velocity of the electromagnetic waves in the vacuum and the phase velocity of the waves in the medium considered: $N = ck/\omega$. If we use the dispersion equation, the optical index of the plasma is then given by:

$$N_{pl} = \sqrt{1 - \frac{\omega_{pe}^2}{\omega^2}}$$

The phase velocity and the group velocity of the electromagnetic wave in the plasma can also be calculated explicitly:

$$v_{\varphi} = \frac{c}{\sqrt{1 - \dfrac{\omega_{pe}^2}{\omega^2}}} \ ; \ v_g = \frac{\partial \omega}{\partial k} = c\sqrt{1 - \frac{\omega_{pe}^2}{\omega^2}}$$

The optical index of the plasma is real only if $\omega > \omega_{pe}$. In this case, the electromagnetic wave can propagate in the plasma, but its optical index is smaller than 1, contrary to what happens in the usual dielectrics. This is because the phase velocity of an electromagnetic wave in plasma is larger than the speed of light. The result is that a light beam that enters plasma will move away from the normal to the air plasma interface (the opposite of what happens when it enters water or glass). A plasma lens with thin edges will be divergent and a plasma lens with thick edges will converge. Note that the group velocity, which is related to the phase velocity by $v_g v_\phi = c^2$, is always smaller than the speed of light. As the transport of energy in a wave packet is associated with the group speed, we respect the constraints of relativity!

The frequency limit $\omega = \omega_{pe}$ is called a *cutoff frequency*. The general definition of a cutoff frequency is as follows: if N (or k) tends to zero for a finite value of the frequency, this frequency is called a cutoff frequency. As the index N_{pl} and the wave vector k tend towards 0, if the frequency tends towards the plasma frequency, the latter is indeed a cutoff frequency for the electromagnetic waves. If $\omega < \omega_{pe}$, the wave vector becomes imaginary, and we can define an evanescence length or penetration length of the wave in the plasma:

$$L_p = \frac{c}{\omega_{pe}\sqrt{1 - \omega^2/\omega_{pe}^2}}$$

Waves with frequency lower than the *cutoff frequency do not propagate in the plasma*. Consequently, if we consider a wave of a lower frequency than the plasma frequency $\omega < \omega_{pe}$, which arrives on an air–plasma interface, it will be reflected by the plasma.

If a wave is incident with an angle θ_0 with respect to the normal to the surface of the plasma (see figure 5.4) with a frequency such that $\omega_{pe} < \omega < \omega_{pe}/\cos\theta_0$, it will also be reflected. This result can be easily shown using geometrical optics and assimilating the optical index of air to that of vacuum ($N = 1$).

The frequency and the component of the wave vector parallel to the surface k_y (Snell's law) are constant as the wave propagates across the surface; we can then write, using the dispersion equation:

$$k_z^2 = k^2 - k_y^2 = \frac{\omega^2 - \omega_{pe}^2}{c^2} - \left(\frac{\omega}{c}\sin\theta_0\right)^2$$

that is:

$$k_z = \frac{1}{c}\sqrt{\omega^2 \cos^2 \theta_0 - \omega_{pe}^2}.$$

The condition for the wave vector to be real is therefore: $\omega > \omega_{pe}/\cos\theta_0$.

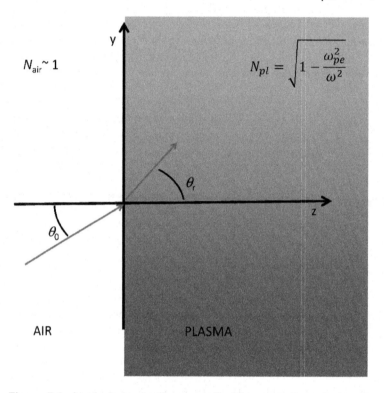

Figure 5.4. *An electromagnetic wave with a frequency $\omega\cos\theta_0 > \omega_{pe}$ is refracted at the air–plasma interface. The angle of refraction is greater than the incident angle because the plasma index is smaller than 1. For a color version of this figure, see www.iste.co.uk/belmont/plasma.zip*

As a consequence, an electromagnetic wave of given frequency, which passes through a plasma of increasing density in the direction of the density gradient, can propagate only up to the cutoff frequency $\omega_{pe} = \omega$. Similarly, an electromagnetic wave of given frequency, which passes through a plasma of increasing density at an oblique incident with an angle θ_0 with respect to the density gradient, can only propagate to the point where $\omega_{pe} = \omega\cos\theta_0$.

The ionosphere, for example, prevents the reception of electromagnetic waves from space which have a frequency lower than 6-10 MHz (plasma frequency corresponding to the maximum electron density of the ionosphere) on the ground, or reflects the waves of the same frequency transmitted on Earth, as shown in Figure 5.5. Here we present a simplified model of the ionosphere with colors increasingly darker as the plasma density increases. In this diagram, the component of the wave vector of the electromagnetic wave perpendicular to the interface varies according to the dispersion equation:

$$k_z = \frac{\omega}{c}\sqrt{\cos^2 \theta_0 - \frac{\omega_{pe}^2}{\omega^2}}$$

The quantity ω_{pe} increases from one layer to the next. At the beginning $\omega \cos \theta_0 > \omega_{pe}$, but, as the wave propagates, ω_{pe} *becomes* closer to $\omega \cos\theta_0$ so k_z decreases (and the wavelength λ increases, as we see in Figure 5.5). If k_z decreases, the refractive index N_{pl} decreases and the direction of propagation changes, because the wave vector is moving further and further away from the normal to the interface according to Snell's law (see Figure 5.5). This continues until the wave vector is zero and the wave is reflected. We can also note on the figure that the amplitude of the electric field of the wave increases; this is related to the conservation of energy[4].

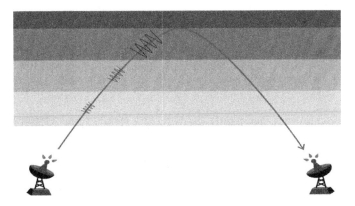

Figure 5.5. *The electric field of an electromagnetic wave of frequency ω, which is reflected by the lower layers of the ionosphere. The wave propagates if $\omega \cos\theta_0 > \omega_{pe}$, but, as the plasma density (and therefore ω_{pe}) increases, from a certain point $\omega_{pe} = \omega\cos\theta_0$ and the wave is reflected. For a color version of this figure, see www.iste.co.uk/belmont/plasma.zip*

---

4 The conservation of energy is reflected in the fact that the signal intensity, that is, the average of the Poynting vector, $\langle S \rangle$, is constant. Since $\langle S \rangle = \varepsilon_0 v_g |E|^2 = \varepsilon_0 c\ N_{pl}|E|^2 =$ cste, the electric field increases as the optical index of the plasma decreases.

This principle was used at the beginning of ionospheric exploration to study the position and density of the ionosphere. The ionospheric probing made it possible to obtain ionograms like the one presented in Figure 5.6 and to deduce the density profile of the bottom of the ionosphere.

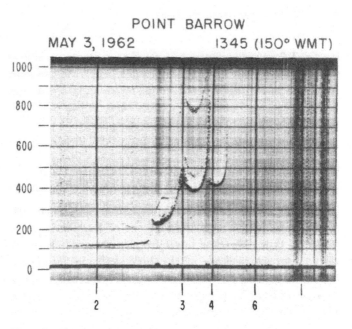

Figure 5.6. *Ionogram (frequency in abscissa in MHz, reflection altitude in ordinate in km) (source: Davies, 1966)*

The same phenomenon exists in metals: the cutoff frequency of electromagnetic waves explains their brightness[5].

Finally, we see that the initial hypothesis that the ions are immobile is satisfied, because the condition of existence of these waves is that the frequency is greater than the electronic plasma frequency, corresponding to a very fast movement with respect to the reaction time of the ions which have a large inertia.

---

5 Indeed, as far as the interaction with light is concerned, we can treat the conduction electrons of the metal as a cold plasma whose plasma frequency is $\omega_{pe} \approx (20\pi - 80\pi)10^{14}$ rad/s. The frequencies of the visible radiation are in the interval $\omega_{vis} \approx (8\pi - 15\pi) \ 10^{14}$ rad/s : as they are lower than the plasma frequency of the free electrons of the metal, the visible light is reflected and the metal shines.

5.5.3. "Electronic" waves in a magnetized plasma

We will resume the calculation here in a more general way, without making any hypothesis on the electrostatic or electromagnetic nature of the waves, and by extending the calculation to the case of a non-zero static magnetic field $\mathbf{B}_0$. The waves that we want to study here are always "high-frequency" modes and therefore we can consider the ions as a neutralizing background, without solving the ions equations.

The fluid equations used are identical to those studied in the previous section, and this time the term $\mathbf{u}_e \times \mathbf{B}_0$ introduced in the impulse transport equation gives a contribution to order 1, because now we have an equilibrium magnetic field $\mathbf{B}_0$ and the term $\mathbf{u}_{e1} \times \mathbf{B}_0$ is not negligible. Hence, instead of finding a dispersion equation like equation [5.18], we will find several dispersion equations, associated with different polarizations of the waves, which make it possible to cancel the determinant of the system of equations.

It is convenient, to explain the result and study the properties of the solutions, to choose a reference frame where the wave vector is along x and where $\mathbf{B}_0$ is in the x,z plane: this makes it easy to distinguish the solutions having polarization with $\mathbf{E}$ parallel to $\mathbf{k}$ (longitudinal or electrostatic component) and those for which $\mathbf{E}$ is perpendicular to $\mathbf{k}$ (transverse components). We then find, by introducing the angle θ between $\mathbf{k}$ and $\mathbf{B}_0 = B_0\,\mathbf{b}$:

$$\mathbf{k} = \begin{bmatrix} k \\ 0 \\ 0 \end{bmatrix} \text{ and } \mathbf{b} = \begin{bmatrix} \cos\theta \\ 0 \\ \sin\theta \end{bmatrix}$$

In this situation, it is easier to explicitly write the matrix corresponding to the linearized homogeneous system for exponential solutions for a monochromatic wave. This will have the form $\mathbf{D} \cdot \mathbf{E}_1 = 0$:

$$\Rightarrow \mathbf{D} = \begin{bmatrix} \omega^2 - \omega_{pe}^2 - \gamma k^2 V_{the}^2 & -i\dfrac{\omega^2 - k^2 c^2}{\omega}\omega_{ce}\sin\theta & 0 \\[2ex] i\omega\omega_{ce}\sin\theta & \omega^2 - \omega_{pe}^2 - k^2 c^2 & -i\dfrac{\omega^2 - k^2 c^2}{\omega}\omega_{ce}\cos\theta \\[2ex] 0 & i\dfrac{\omega^2 - k^2 c^2}{\omega}\omega_{ce}\cos\theta & \omega^2 - \omega_{pe}^2 - k^2 c^2 \end{bmatrix}$$

The general dispersion equation is obtained by canceling the determinant of this matrix.

In the absence of a magnetic field, the four non-diagonal terms are null; the resolution is very simple. There are three modes, two of which are identical, corresponding to the cancellation of each of the three diagonal terms: these are the modes that were calculated in the first part.

– *The two identical solutions* have a dispersion equation $\omega^2 = \omega_{pe}^2 + k^2 c^2$, and they have transverse polarizations (nothing distinguishes the transverse directions y and z in the absence of B_0). These solutions correspond, as was foreseeable, to the propagation of electromagnetic waves in plasma. Note that even if the plasma is hot, there is no contribution of the pressure to transverse modes; the dispersion relation is the same as for the cold plasma.

– *The third mode* has for dispersion equation $\omega^2 = \omega_{pe}^2 + 3k^2 V_{the}^2$ and it has a strictly longitudinal polarization: it is thus exactly the electrostatic mode that had provided the computation in section 5.3. We see that this mode is not a modification of the vacuum modes but a distinct and additional solution.

5.5.4. *Role of the static magnetic field*

In the presence of a static magnetic field, that is, for non-zero ω_{ce}, we see that the three above-mentioned modes are "coupled". Their dispersions and polarizations become hybrid: the initially electrostatic mode acquires a transverse component and vice versa. We can briefly look at the case where the waves propagate perpendicular to the magnetic field. In this case, $\cos\theta = 0$ and $\sin\theta = 1$. A solution that makes the determinant null is the usual solution for transverse modes, $\omega^2 = \omega_{pe}^2 + k^2 c^2$. This mode is called the ordinary mode. The other solutions correspond to the condition:

$$\frac{c^2 k^2}{\omega^2} = 1 - \frac{\omega_{pe}^2}{\omega^2} \frac{\omega^2 - \omega_{pe}^2}{\omega^2 - (\omega_{pe}^2 + \omega_{ce}^2)}$$

For simplicity, the effects of electronic temperature have been neglected and $V_{the} = 0$. These solutions are called extraordinary modes. From this equation, we can see that, for $\omega = \sqrt{\omega_{pe}^2 + \omega_{ce}^2} \equiv \omega_H$, the wave vector k diverges. This frequency is called the *upper hybrid frequency*. This frequency is a *resonant frequency*: a frequency is called a resonance when there is a finite value of the frequency, such that, for this value, the wave vector (or the optical index) tends towards infinity. If we are close to a resonance, the waves will be absorbed by the medium. We can also calculate the cutoff frequencies (frequencies for which the wave vector becomes zero, as in the case of an electromagnetic wave in non-magnetized plasma). Two cutoff frequencies are obtained:

$$\omega_D = \frac{1}{2}\left[\omega_{ce} + \sqrt{\omega_{ce}^2 + 4\omega_{pe}^2}\right]$$

$$\omega_G = \frac{1}{2}\left[-\omega_{ce} + \sqrt{\omega_{ce}^2 + 4\omega_{pe}^2}\right]$$

The extraordinary wave only propagates if $\omega_G < \omega < \omega_H$; $\omega > \omega_D$.

In the case of propagation parallel to the magnetic field ($\cos\theta = 1$ and $\sin\theta = 0$), we have the same cutoff frequencies, which correspond to two modes: one that propagates with a circular polarization to the right, and the other with circular polarization to the left.

The two transverse modes, which were identical in the absence of a magnetic field, are distinguishable from each other (*ordinary* and *extraordinary*) if there is an equilibrium magnetic field. All these modifications remain weak if $\omega_{ce} \ll \omega_{pe}$, but they are maximum at the neighborhood of the intersection of the modes, that is, for small values of k, close to $\omega = \omega_{pe}$ (figure 5.7). It can be seen in the figure that the plasma oscillation is noticeably modified by the presence of $\mathbf{B}_0$ only for small values of k. In this limit, the phase velocity is of the order of the speed of light.

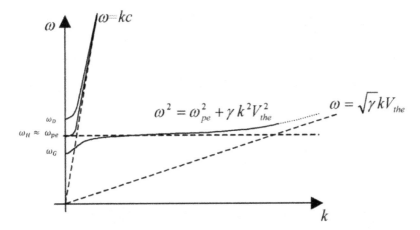

Figure 5.7. *Fluid dispersion equation in the presence of a weak magnetic field taking into account the effects of finite electronic temperature. For the longitudinal or quasi-longitudinal mode if the temperature tends towards 0, the asymptote indicates the presence of a resonance*

5.6. Two-fluid system: low-frequency propagation modes

In plasma, we can also excite low-frequency modes, for which not only the electron motion is important, but also the movement of ions. We will see two examples: ion-acoustic waves in non-magnetized plasma, and transverse Alfvén waves in magnetized plasma.

5.6.1. Ion-acoustic waves

Let us consider a non-magnetized, homogeneous hot plasma at rest. Its equilibrium is defined by:

$$Zn_{i0} = n_{e0}, \mathbf{u}_{e0} = \mathbf{u}_{i0} = 0, T_e = T_{e0}, T_i = T_{i0}, \mathbf{E}_0 = 0, \mathbf{B}_0 = 0$$

The system of equations to analyze is:

$$\begin{cases} d_t(n_e) + n_e \nabla.(\mathbf{u}_e) = 0 \\ n_e m_e d_t(\mathbf{u}_e) + \nabla(p_e) = -n_e e(\mathbf{E} + \mathbf{u}_e \times \mathbf{B}) \\ d_t\left(\dfrac{p_e}{n^{\gamma_e}_e}\right) = 0 \end{cases}$$

$$\begin{cases} d_t(n_i) + n_i \nabla.(\mathbf{u}_i) = 0 \\ n_i m_i d_t(\mathbf{u}_i) + \nabla(p_i) = Zn_i e(\mathbf{E} + \mathbf{u}_i \times \mathbf{B}) \\ d_t\left(\dfrac{p_i}{n^{\gamma_i}_i}\right) = 0 \end{cases}$$

$$\nabla.(\mathbf{E}) = -e\frac{n_e - Zn_i}{\varepsilon_0}$$

Notice that two different indices $\gamma_{i,e}$ have been used, because the electrons and the ions will not respond in the same way to the solicitations related to the wave's propagation. Regarding the ions, the index is related to fast compressions and decompressions on their time scale; they do not have time to exchange energy, and therefore, they behave adiabatically in one dimension: in this case $\gamma_i = 3$. As far as the electrons are concerned, the typical time of compressions and decompressions is very slow with respect to their reaction time scale; they therefore have the time to exchange energy and smooth temperature gradients. As a consequence, they have isothermal behavior, that is, $\gamma_e = 1$ (constant temperature). All this can be verified *a posteriori*, by comparing the phase velocity of the wave with the thermal velocity of the particles for each species: the phase velocity is low compared with the thermal velocity of the electrons, and larger or of the same order as the thermal velocity of the ions. The linearized equations, for plane wave solutions of the type $e^{i(kx-\omega t)}$, have the form:

$$\begin{cases} -i\omega\, n_{e1} + n_{e0} i\mathbf{k} \cdot \mathbf{u}_{e1} = 0 \\ -i\omega\, n_{e0} m_e \mathbf{u}_{e1} + ik_B T_{e0} \mathbf{k}\, n_{e1} = -n_{e0} e\mathbf{E}_1 \end{cases}$$

$$\begin{cases} -i\omega n_{i1} + n_{i0} i\mathbf{k} \cdot \mathbf{u}_{i1} = 0 \\ -i\omega n_{i0} m_i \mathbf{u}_{i1} + 3ik_B T_{i0} \mathbf{k}\, n_{i1} = Zn_{i0} e\mathbf{E}_1 \end{cases}$$

$$i\mathbf{k} \cdot \mathbf{E} = -e\frac{n_{e1} - Zn_{i1}}{\varepsilon_0}$$

Before solving this algebraic system, we can make a simplification and neglect the inertia of the electrons in the momentum equation. This corresponds to neglecting the term $\omega n_{e0} m_e \mathbf{u}_{e1}$ with respect to $k_B T_{e0} k n_{e1}$, and we can once again verify *a posteriori* that this hypothesis is satisfied because the phase velocity of the wave is small compared to the thermal velocity of the electrons. The solution of this system gives the dispersion relation of the ion-acoustic waves.

$$\omega^2 = \frac{k^2 c_s^2}{1 + k^2 \lambda_{De}^2} + k^2 c_s^2 \left(\frac{3T_{i0}}{ZT_{e0}} \right) \qquad [5.19]$$

Here $c_s^2 = Zk_B T_{e0}/m_i$ is the ion-acoustic speed, which depends on the electronic temperature and the ionic inertia. The velocity of propagation of the wave is mainly determined by this speed. The second term is negligible compared to the first in all experimental situations, where $T_{i0} \ll T_{e0}$. Moreover, as we will see in Chapter 6, the condition of existence of these waves imposes $ZT_{e0}/T_{i0} \gg 1$. In the opposite case, the complete kinetic calculation will show that the only solutions that exist are damped waves ("Landau damping"), that is, solutions in $e^{-i\omega t - \gamma t}$.

If we compare the phase velocity of the wave with the electronic thermal velocity, we see that the ratio is equal to sqrt(Zm_e/m_i) and therefore the phase velocity of the wave is much smaller than the thermal velocity: consequently, we verify *a posteriori* that the hypothesis that was used to neglect the inertia of the electrons and to arrive at equation [5.19] is satisfied.

In addition, for long wavelengths, we can neglect $k^2 \lambda_{De}^2$ in the denominator with respect to 1. This is equivalent to considering the hypothesis of quasi-neutrality, that is, $n_{el} \approx Z n_{i1}$, and neglecting the electric field $\mathbf{E}_1$ in the Maxwell–Gauss equation. It is important to point out that it does not mean that the electric field is zero; it is just small and must be kept in the other equations, because it is precisely such a field that links the ionic and electronic movements.

Finally, we observe the phenomenon of resonance, which has already been introduced for electromagnetic waves in the presence of an external magnetic field. If we neglect the last term of equation [5.19], which must be small for the wave to exist as discussed previously, we can calculate k from the dispersion equation, and we obtain $k^2 = \dfrac{1}{\lambda_{De}^2} \dfrac{\omega^2}{\omega_{pi}^2 - \omega^2}$, where we used the relation $c_s = \lambda_{De} \omega_{pi}$, with $\omega_{pi}^2 = \dfrac{n_{io} Z^2 e^2}{\varepsilon_0 m_i}$. From this equation, we can see that if $\omega = \omega_{pi}$, k tends to infinity. This is the definition of a resonance frequency, when there is a finite value of the frequency, such that, for this value, the wave vector (or optical index) tends to infinity. In general, $\omega < \omega_{pi}$. If we are close to a resonance, the waves will be absorbed by the medium, and if we have a system with decreasing density, the frequency will remain the same, but the wavelengths will decrease (and the wave vectors increase), until the waves are absorbed.

5.6.2. *Alfvén Waves*

Alfvén waves[6] are low-frequency electromagnetic waves that can be excited in magnetized plasma at rest. This is a particular – and particularly important – example of MHD waves that we will study more generally in the next section. In this case, we can neglect the effects of temperature, because they do not participate in the propagation of these waves. The equilibrium is given by (Figure 5.8):

$$Z n_{i0} = n_{e0}, \ \mathbf{u}_{e0} = \mathbf{u}_{i0} = 0, \ T_e = 0, \ T_i = 0, \ \mathbf{E}_0 = 0, \ \mathbf{B}_0 = B_0 \mathbf{e}_z$$

---

6 Also called torsional Alfvén waves to differentiate them from other low-frequency modes in magnetized plasma that are called compressional Alfvén waves or magneto-sonic waves. These compressional modes will be studied in the next section.

We consider waves propagating in the direction of the equilibrium magnetic field that is, in the z direction. The plane wave solutions are thus of the form $e^{i(kz-\omega t)}$. It is also assumed that the solutions are transverse modes, that is, such that the direction of propagation, the electric field and the magnetic field are perpendicular with, for example, $\mathbf{F}_1 = F_1\mathbf{e}_x$ and $\mathbf{B}_1 = B_1\mathbf{e}_y$. As we have already done for electromagnetic waves in an unmagnetized plasma, we can combine the Maxwell–Ampere equation with the Maxwell–Faraday equation and replace the magnetic field variable. Notice that for this mode it is necessary to keep the electrons and ions contribution to the current. The resulting linear equation (wave propagation equation) is:

$$\partial_t^2(\mathbf{E}_1) - c^2\Delta(\mathbf{E}_1) = \frac{1}{\varepsilon_0}\partial_t(\mathbf{j}_1) = \frac{1}{\varepsilon_0}\partial_t(Zen_{i0}\mathbf{u}_{i1} - en_{e0}\mathbf{u}_{e1})$$

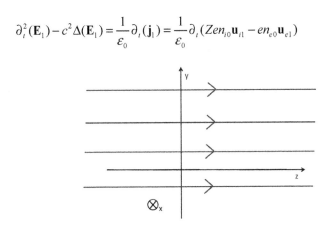

Figure 5.8. B_0 equilibrium magnetic field for Alfvén waves. For a color version of this figure, see www.iste.co.uk/belmont/plasma.zip

The mass conservation equations give no useful information because the density perturbation for these waves is zero (as for the other transverse electromagnetic waves). On the other hand, the momentum conservation equations for the ions and the electrons, linearized, are a little more complicated than what we have already seen, because of the presence of the equilibrium magnetic field. Projected on the axes, they take the form:

$$\begin{cases} -i\omega\, m_e u_{e1,x} = -e(E_1 + u_{e1,y}B_0) \\ -i\omega\, m_e u_{e1,y} = eu_{e1,x}B_0 \end{cases}$$

$$\begin{cases} -i\omega\, m_i u_{i1,x} = Ze(E_1 + u_{i1,y}B_0) \\ -i\omega\, m_i u_{i1,y} = -Zeu_{i1,x}B_0 \end{cases}$$

This system can be easily solved to obtain the velocities as a function of the electric field. This will allow us to have the first-order current to replace in the wave propagation equation. We obtain:

$$\begin{cases} u_{e1,x} = -i\dfrac{e}{m_e\,\omega}\dfrac{E_1}{\left(1-\dfrac{\omega_{ce}^2}{\omega^2}\right)} \approx 0 \\[3em] u_{e1,y} = \dfrac{e}{m_e\,\omega}\dfrac{|\omega_{ce}|}{\omega}\dfrac{E_1}{\left(1-\dfrac{\omega_{ce}^2}{\omega^2}\right)} \approx -\dfrac{E_1}{B_0} \end{cases}$$

[5.20]

$$\begin{cases} u_{i1,x} = i\dfrac{Ze}{m_i\,\omega}\dfrac{E_1}{\left(1-\dfrac{\omega_{ci}^2}{\omega^2}\right)} \approx -i\dfrac{Ze}{m_i\,\omega}\dfrac{\omega^2}{\omega_{ci}^2}E_1 \\[3em] u_{i1,y} = \dfrac{e}{m_i\,\omega}\dfrac{\omega_{ci}}{\omega}\dfrac{E_1}{\left(1-\dfrac{\omega_{ci}^2}{\omega^2}\right)} \approx -\dfrac{E_1}{B_0} \end{cases}$$

[5.21]

The electron and ion cyclotron frequencies, ω_{ce} and ω_{ci}, have been introduced in Chapter 2. By hypothesis, we study low-frequency modes; we therefore have $\omega \ll \omega_{ci} \ll \omega_{ce}$ leading to the approximate equalities for the right-hand side of equations [5.20] and [5.21]. We see immediately that the electrons and the ions move together in the y direction and, therefore, there is a movement, but no current, because the charges are compensated exactly. This movement is due to the drift velocity $\mathbf{V}_m = \dfrac{\mathbf{E}}{B}\times\mathbf{b}$ introduced in Chapter 2. On the other hand, there is a current in the x direction, which is due to the ionic movement. Therefore, the propagation equation of the electromagnetic wave takes the form:

$$\omega^2 E_1 - c^2 k^2 E_1 \approx i\omega\frac{en_{e0}}{\varepsilon_0}n_{i0}u_{i1,x}$$

If we combine this with the equation for u_{i1x}, we obtain the dispersion equation for Alfvén waves:

$$\boxed{\omega^2 = \dfrac{k^2 V_A^2}{1+\dfrac{V_A^2}{c^2}}}$$

[5.22]

where we introduced the Alfvén velocity $V_A = \sqrt{\dfrac{B_0^2}{m_i n_{i0} \mu_0}}$. This mode was calculated,

for the first time, by H. Alfvén (Nobel Prize, 1942). In general, the Alfvén velocity is much smaller than the speed of light $V_A \ll c$, and we have $\omega \approx k V_A$. In the general case of oblique propagation, the dispersion equation becomes $\omega = k_{//} V_A$. It is interesting to try to visualize these waves; if we draw the field lines of the Alfvén waves, they behave like waves that propagate on a rigid string, where the field line takes the place of the string (Figure 5.9).

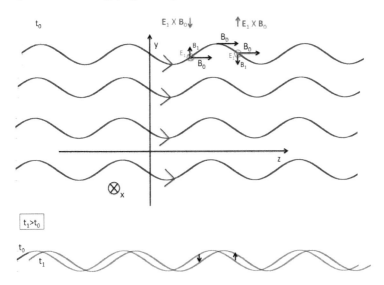

Figure 5.9. *Alfvén waves, the magnetic and electric field perturbation are deliberately exaggerated in size for the sake of the illustration. At the top, we show an image of the field lines at a given time; at the bottom, we show a field line at two successive times (propagation to the right). For a color version of this figure, see www.iste.co.uk/belmont/plasma.zip*

In this figure, we have explicitly drawn the directions of B_1, E_1 and of the drift velocity $\mathbf{E}_1 \times \mathbf{B}_0$ in two positions in z on the first field line. We see that the plasma drift velocity is negative for the first point, and positive for the second point. If we look at how the field line moves in the lower figure at two successive times, we see that it moves in the same way. In other words, the line moves down or up with

velocity $v_l = \dfrac{\omega}{k} \left| \dfrac{B_1}{B_0} \right| = \left| \dfrac{E_1}{B_0} \right|$, where the second equality comes from the Maxwell–

Faraday equation. The field line is then modified by the propagation of the wave in the same way as the plasma, as if the field line was frozen in the plasma (no relative movement). This is an example of the application of the frozen-in law of magnetic field lines, already introduced in Chapter 4. If these waves are excited in a magnetized cylinder, with the equilibrium field in the axial direction, the electric field will be radial, and the magnetic field perturbation and the drifting motion will be azimuthal, which produces a twisting of the field lines.

Alfvén mode guidance: the Alfvén mode has a frequency that depends only on the component of the wave vector that is parallel to the equilibrium field, $k_{//}$. The group velocity is therefore parallel to $\mathbf{B_0}$: it is said to be a guided mode. This plays a great role in plasmas because a perturbation in this mode can propagate very far from its source: this mode is not reduced by a geometric factor $1/r^2$ as modes that propagate isotropically such as sound waves. The presence, for example, of a conducting body in magnetized plasma in motion corresponds to a perturbation of this plasma that can be detected extremely far[7].

5.7. MHD propagation modes

We have seen in Chapter 4 that we can describe a magnetized plasma by the MHD system of equations, that is, the system is treated as a single fluid, but where, for example, the speed and the current are independent variables (unlike the two-fluid system, where both are expressed as a function of the velocities of the two species, ions and electrons). We will now apply to the MHD equations the general technique for finding eigenmodes for the linearized system. It is expected to find in this way the low-frequency modes in magnetized plasma that have been studied in the previous section since this is the limit where the MHD equations are valid (see Chapter 4). Moreover, there are other possible modes of propagation that we have not studied yet.

5.7.1. *Writing and resolution of the linearized system*

We consider a plasma that can be described by the MHD equations system. Moreover, we assume that the closure equation (connecting $p = p_e + p_i$ to $n = n_e$ or Zn_i) is polytropic law $d_t \left(p / n^\gamma \right) = 0$. Let us consider the case of an equilibrium ("zeroth order"), uniform and stationary, with $u_0 = 0$ and $E_0 = 0$. If we apply the first

---

7 Behind a conductor that moves in a magnetic field (e.g. a satellite – natural or artificial – in the magnetosphere of a planet), we show that there is a kind of "magnetic wake" in the form of two long cylinders inclined with respect to the direction of the movement called "Alfvén wings".

three points of the technique of section (5.2), the operations of linearization and algebrization give:

$$\partial_t (n_1) + n_0 \nabla.(\mathbf{u}_1) = 0$$

$$n_0 m \partial_t (\mathbf{u}_1) + \nabla(p_1) = \frac{1}{\mu_0} (\nabla \times \mathbf{B}_1) \times \mathbf{B}_0$$

$$\frac{p_1}{p_0} = \gamma \frac{n_1}{n_0}$$

$$\nabla \times (\mathbf{u}_1 \times \mathbf{B}_0) = \partial_t (\mathbf{B}_1)$$

which means that:

$$\left\{ \begin{array}{l} \dfrac{n_1}{n_0} = \mathbf{k}.\dfrac{\mathbf{u}_1}{\omega} \\[3mm] -\omega \mathbf{u}_1 + \mathbf{k}\,\dfrac{p_1}{n_0 m} = V_A^2 \left(k_{//} \dfrac{\mathbf{B}_1}{B_0} - \mathbf{k}\,\dfrac{\mathbf{B}_1 \cdot \mathbf{B}_0}{B_0^2} \right) \\[3mm] \dfrac{p_1}{n_0 m} = c_s^2 \dfrac{n_1}{n_0} \\[3mm] \dfrac{\mathbf{B}_1}{B_0} = -k_{//} \dfrac{\mathbf{u}_1}{\omega} + \dfrac{\mathbf{B}_0}{B_0}\dfrac{(\mathbf{k} \cdot \mathbf{u}_1)}{\omega} \end{array} \right.$$

We did not rewrite the last equation $\nabla.(\mathbf{B}) = 0$; it was directly included in the linearized equations by considering $\mathbf{k}.\mathbf{B}_1 = 0$. Here, $k_{//} = \mathbf{k}.\mathbf{b}$ where we defined $\mathbf{b} = \mathbf{B}_0/B_0$; $V_A = \sqrt{\dfrac{B_0^2}{m n_0 \mu_0}}$ and the speed of sound is defined by $c_s^2 = \dfrac{\gamma p_0}{n_0 m} = k_B \dfrac{\gamma_e Z T_e + \gamma_i T_i}{m_i} = k_B \dfrac{Z T_e + 3 T_i}{m_i}$. To calculate the speed of sound, we used the usual MHD hypothesis, $m_e \ll m_i$ and $n_0 m \approx Z n_{0i} m_i$.

We can now solve this linear system by first replacing the scalars n_1 and p_1. We thus obtain two equations for the vectors $\mathbf{u}_1$ and $\mathbf{B}_1$, and with some manipulations, we can reduce the system to three equations (the three components of the vector $\mathbf{u}_1$), and write it in the form $\mathbf{D}.\mathbf{u}_1 = 0$ characteristic of a homogeneous system. To do this, it is convenient now to specify the direction of the equilibrium magnetic field and the wave vector. Let us choose the reference where:

$$\mathbf{b} = \begin{bmatrix} 0 \\ 0 \\ 1 \end{bmatrix} \text{ and } \mathbf{k} = \begin{bmatrix} k_\perp \\ 0 \\ k_{//} \end{bmatrix}$$

which gives *the propagation matrix* $\mathbf{D}$ (matrix of the coefficients for the resolution in u_1):

$$\mathbf{D} = \begin{bmatrix} \omega^2 - k_\perp^2 c_s^2 - k^2 V_A^2 & 0 & -k_{//} k_\perp c_s^2 \\ 0 & \omega^2 - k_{//}^2 V_A^2 & 0 \\ -k_{//} k_\perp c_s^2 & 0 & \omega^2 - k_{//}^2 c_s^2 \end{bmatrix}$$

Non-trivial solutions of $\mathbf{D}.\mathbf{u}1 = 0$ necessarily satisfy *the dispersion equation*:

$$\det(\mathbf{D}) = 0 \Leftrightarrow \left(\omega^2 - k_{\parallel}^2 V_A^2\right)\left[\left(\omega^2 - k_\perp^2 c_s^2 - k^2 V_A^2\right)\left(\omega^2 - k_{\parallel}^2 c_s^2\right) - k_{\parallel}^2 k_\perp^2 c_s^4\right] = 0$$

This equation has three pairs of solutions, that is, three "modes" (the equation is bi-square because of the two directions of propagation of each mode).

5.7.1.1. *Alfvén mode*

This is the mode we studied in the previous section, corresponding to the roots of:

$$\boxed{\omega^2 - k_{//}^2 V_A^2 = 0}$$

Its polarization is simple (Figure 5.10): the perturbation of velocity is polarized according to the direction y of the reference, that is, perpendicular to $\mathbf{B_0}$ and $\mathbf{k}$. Since the velocity rotational is non-zero ($\mathbf{k} \times \mathbf{u}_1 \neq 0$), but the divergence is zero ($\mathbf{k}.\mathbf{u}_1 = 0$), this mode is called *torsional*, and there is no compression associated with the movement. It follows that the density perturbation is zero: the flow lines "twist" but do not get closer to each other[8]. The disturbance $\mathbf{B}_1$ of the magnetic field is also polarized along y: the magnetic field rotates in the plane $y0z$, but does not change module.

---

8 Such twists are also possible in a perfect neutral gas ($V_A = 0$), but, in the absence of viscosity, they do not propagate ($\omega = 0$).

The complete nonlinear calculation shows that there exists a solution called "rotational discontinuity" (see Chapter 7), which respects B = cst and which is not fully polarized along y. This corresponds to a ripple of the field lines as in Figure 5.9. In comparison with the calculation of the Alfvén mode in the two-fluid model, it is recalled that in the MHD equations, it is the overall velocity of the plasma system considered as a single fluid that must be taken into account, and the current does not appear explicitly. In addition, the displacement current is neglected in the Maxwell equations in the MHD limit. This explains the difference between the dispersion equation found here and formula [5.22]: the MHD calculation is valid in the limit $V_A \ll c$.

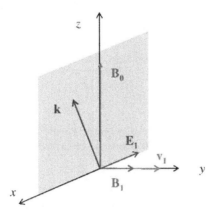

Figure 5.10. *Polarization of the Alfvén mode. For a color version of this figure, see www.iste.co.uk/belmont/plasma.zip*

5.7.1.2. Magneto-sonic modes (fast and slow)

These two modes are given by the solutions of the equation of fourth degree:

$$\left(\frac{\omega}{k}\right)^4 - (V_A^2 + c_s^2)\left(\frac{\omega}{k}\right)^2 + V_A^2 c_s^2 \cos^2\theta = 0 \quad (\text{with } \cos\theta = \frac{k_{//}}{k})$$

We call fast mode the solution which has the fastest phase speed and the other one is called slow. Both modes have, like the Alfvén mode, a rectilinear polarization, but this time, the perturbation $\mathbf{u}_1$ is in the xz plane, that is, in the so-called "co-planarity" plane, containing at the same time the magnetic field $\mathbf{B}_0$ and the wave normal $\mathbf{k}$ (Figure 5.11). These modes are therefore usually compressional ($\mathbf{k}.\mathbf{u}_1 \neq 0$). The polarization of the magnetic field $\mathbf{B}_1$ is also in this same plane, and perpendicular to $\mathbf{k}$. This means that the field lines are deviated (refracted) in the co-planarity plane defined above, but that they do not turn out of this plane. The module of **B**, in general, is not constant: it increases together with the density for the fast mode, in opposition of phase for the slow mode. If the magnetic field is zero, the ion-acoustic mode calculated in section 5.6.1 is recovered, but in the MHD limit, that is, $k\lambda_{de} \ll 1$.

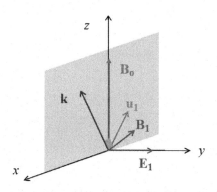

Figure 5.11. *Polarization of magneto-sonic modes. For a color version of this figure, see www.iste.co.uk/belmont/plasma.zip*

5.7.2. *Some remarks*

– Kinetic pressure and magnetic pressure: the β parameter. We can see that, by the calculation of the MHD propagation modes, the thermal effects, which intervene via c_S, can be dominant or negligible with respect to the magnetic effects, which intervene via V_A. This is one of the aspects of the "mechanical/electromagnetic" coupling presented in the introduction. The quantity (already discussed in Chapter 4):

$$\beta = \frac{p}{B^2/2\mu_o} = \frac{2}{\gamma}\frac{c_s^2}{V_A^2}$$

is an important parameter of the MHD since it indicates which of the two terms, "kinetic pressure" or "magnetic pressure", dominates for the plasma dynamics and its frozen field. The "competition" between these two terms can be seen directly in the plasma momentum conservation equation.

– Special case of perpendicular propagation: in the case $k_{//} = 0$, only the fast magneto-sonic mode can propagate at a non-zero speed. Its velocity perturbation $\mathbf{u}_1$ is then polarized along x, that is, parallel to **k**: it therefore has no torsional component. The other two modes (slow mode and Alfvén mode) instead correspond to stationary perturbations with respect to the plasma. In this degenerate case, any combination of the polarizations of the two modes at $\omega = 0$ becomes possible.

5.8. Excitation of waves in plasma

As we have seen in this chapter, several types of waves can propagate in plasma. Depending on the type of wave we want to study, we can use a different model for plasma (mobile or immobile ions, two-fluid treatment or MHD, etc.). This amounts to making assumptions about the behavior of the plasma under the considered conditions. The corresponding approximations can be made *a priori* and justified *a posteriori*. We can also solve the system in its greater generality and make simplifications on the final result. This second method is more cumbersome from the point of view of calculation, while the first is generally more rapid and effective, but requires a certain intuition/knowledge of the result.

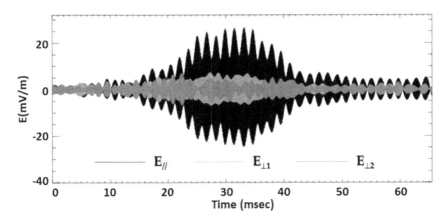

Figure 5.12. *Example of an electrical waveform (amplitude versus time) observed in the interplanetary medium, associated with an eruption on the Sun. The three colors correspond to the three components of the electric field along the interplanetary magnetic field (in black) or perpendicularly (in red and blue). NB: in this figure, the oscillations at the plasma frequency cannot be resolved. The oscillations of the envelope that we see are related to beating phenomenon (source: NASA @ STEREO / WAVE). For a color version of this figure, see www.iste.co.uk/belmont/ plasma.zip*

A significant difference is related to the chosen equilibrium; we do not have the same solutions if the plasma is magnetized or not, or if, at equilibrium, there is a relative velocity between the species. The possibility of exciting a wave in plasma depends on the boundary conditions, the presence of density fluctuations or magnetic field, or the presence of a forcing term. For example, we can excite electromagnetic waves with an antenna and propagate them in plasma.

Electronic plasma waves can be excited by electron beams that pass through plasma. In the laboratory, a charged grid subjected to a pulsed potential in plasma can be used to excite electronic or acoustic-ionic waves, and several waves can coexist at the same time depending on the excitation conditions.

A laser that propagates in plasma can also excite electronic plasma waves or acoustic-ionic waves by nonlinear coupling with the plasma. The universe consists mainly of plasma. It is also an exceptional laboratory to study the development of many waves and instabilities. In particular, they can be studied in detail by *in-situ* measurements of particles and fields (electrical and magnetic) in the planetary environment and the interplanetary medium, as reproduced in Figure 5.12. Plasma waves and acoustic-ionic waves are thus known to be at the origin of the most intense solar electromagnetic emissions emitted in the radio domain (100 kHz–20 MHz).

These waves are associated with electron beams propagating in the interplanetary medium following either an eruption in the upper solar atmosphere or an upstream acceleration of interplanetary shock waves. Plasma waves also have a very important role in the transfer of energy between different spatial scales, and between species (ions and electrons) that cannot otherwise exchange energy given their different mass (and therefore inertia).

Other types of waves are also observed such as Alfvén waves in the ionosphere of the Earth (around 1 Hz), in the interplanetary medium or in the coronal loops of the Sun. In the latter case, they are suspected of participating in the heating of the solar corona, one of the great mysteries of solar physics for over 20 years.

Planets emit electromagnetic radiation in the auroral region. Figure 5.13 illustrates how a spectrogram (time–frequency diagram) of the waves can identify when the satellite making the measurements crosses the radiation emission region.

A summary of the various waves that can propagate in plasma is given below. It includes the examples presented in this chapter. But not only that, we also recall the definition of the cutoff and resonance frequencies.

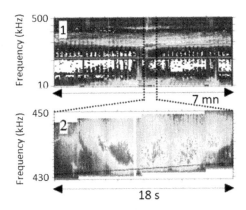

Figure 5.13. *Earth's kilometric radiation (source: R. Pottelette). For a color version of this figure, see www.iste.co.uk/belmont/plasma.zip*

COMMENT ON FIGURE 5.13.– Spectrogram of the waves observed by the Viking (top) and FAST (bottom) satellites in the auroral region (in ordinate the frequency and in abscissa the time); the color code reflects the intensity of the waves. In the upper image (large time window), the radiation spectrum is observed to have a flattened V-shape; the minimum of V coincides with the local cyclotron frequency of the electrons, which suggests that the emission is at this frequency and that, during propagation, the frequency increases. At high temporal resolution (lower image), the spectrum appears to be composed of a large number of fine structures drifting in frequency.

DEFINITIONS.–

– *Cutoff frequency*: we say that a frequency ω is a cutoff frequency if the wave vector associated with this frequency $k(\omega) = 0$. In this case, the optical index of the medium is also equal to zero. A cutoff frequency is generally associated with reflection.

– *Resonant frequency*: a frequency ω is said to be a resonant frequency if the wave vector associated with this frequency $k(\omega) = \infty$. In this case, the optical index of the medium also tends towards infinity. A resonance frequency is generally associated with absorption.

Modes	Plasma motion	Condition	Dispersion equation	Name
Electro-magnetic	Electrons (high frequency)	$\vec{B}_0 = 0$ or $\vec{k} \parallel \vec{B}_0$	$\omega^2 = \omega^2_{pe} + 3k^2 v^2_{the}$	Plasma wave (Langmuir wave)
		$\vec{k} \perp \vec{B}_0$	$\omega^2 = \omega^2_{pe} + \omega^2_{ce} \equiv \omega^2_H$	Hybrid high oscillation
	Electrons and ions (low frequency)	$\vec{B}_0 = 0$ or $\vec{k} \parallel \vec{B}_0$	$\omega^2 = \dfrac{k^2 c^2_{s0}}{1 + k^2 \lambda^2_{de}} + k^2 c^2_{s0}\left(\dfrac{3T_{i0}}{ZT_{e0}}\right); c^2_{s0} \equiv \dfrac{Zk_B T_{e0}}{m_i}$	Acoustic-ionic wave
		$\vec{k} \perp \vec{B}_0$ (almost)	$\omega^2 = \omega^2_{ci} + k^2 c^2_{s0}$	Ionic-electrostatic cyclotron wave
		$\vec{k} \perp \vec{B}_0$ (identically)	$\omega^2 = \dfrac{1}{(\omega_{ci}\omega_{ce})^{-1} + \omega^{-2}_{pi}} \equiv \omega^2_{LH}$	Low hybrid oscillation

Table 5.1. Summary of the main electromagnetic modes in plasma

Modes	Plasma motion	Condition	Dispersion equation	Name
Electro-magnetic	Electrons (high frequency)	$\vec{B}_0 = 0$	$\omega^2 = \omega_{pe}^2 + k^2 c^2$	Light waves
		$\vec{k} \perp \vec{B}_0, \vec{E}_1 \parallel \vec{B}_0$	$\omega^2 = \omega_{pe}^2 + k^2 c^2$	Ordinary waves
		$\vec{k} \perp \vec{B}_0, \vec{E}_1 \perp \vec{B}_0$	$\dfrac{c^2 k^2}{\omega^2} = 1 - \dfrac{\omega_{pe}^2}{\omega^2}\,\dfrac{\omega^2 - \omega_{pe}^2}{\omega^2 - \omega_H^2}$	Extraordinary waves
		$\vec{k} \parallel \vec{B}_0$ (circular right polarization)	$\dfrac{c^2 k^2}{\omega^2} = 1 - \dfrac{\omega_{pe}^2/\omega^2}{1-(\omega_{ce}/\omega)}$	Right mode (whistling wave)
		$\vec{k} \parallel \vec{B}_0$ (circular left polarization)	$\dfrac{c^2 k^2}{\omega^2} = 1 - \dfrac{\omega_{pe}^2/\omega^2}{1+(\omega_{ce}/\omega)}$	Left mode
	Electrons and ions (low frequency)	$\vec{B}_0 = 0$		No solution
		$\vec{k} \parallel \vec{B}_0$	$\omega^2 = \dfrac{k^2 v_A^2}{1 + v_A^2/c^2}$	Torsional Alfvén wave
		$\vec{k} \perp \vec{B}_0$	$\omega^2 = k^2\,\dfrac{c_{s0}^2 + v_A^2}{1 + v_A^2/c^2}$	Compression wave (Magneto-sonic)

Table 5.2. Summary of the main electromagnetic modes in plasma

5.9. Annexes

5.9.1. *CMA Diagram*

This diagram illustrates the variety of possible propagation modes in magnetized plasma and summarizes their properties. It owes its name to its inventors, Clemmow, Mullaly and Allis. Its validity is limited to zero-temperature plasmas, $T_e = T_i = 0$.

It is a tool to find, in given plasma, modes that can propagate. It therefore has a graduated abscissa proportional to the density and an ordinate proportional to the magnetic field. A given frequency wave corresponds to a point in the diagram. The pulsation is normalized with respect to the characteristic pulsations ω_{pe} and ω_{ce} of the electrons. The different areas of the diagram are delimited by cutoff frequencies or resonances. Inside each region of the plane, a small drawing shows the properties of the waves that can propagate. These solutions depend not only on the properties of the plasma, but also on the direction of propagation (Figure 5.14).

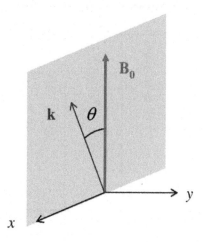

Figure 5.14. *Definition of angle θ. For a color version of this figure, see www.iste.co.uk/belmont/plasma.zip*

The properties of the solution are summarized by a small drawing that shows how the phase velocity varies according to the angle θ. Figure 5.15 shows the principle: we must imagine that the vertical direction is that of the magnetic field; the solution necessarily has a cylindrical symmetry around the direction of the field. But there is not necessarily a solution for all angles of propagation (see Figure 5.16).

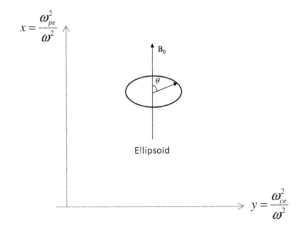

Figure 5.15. *Case where there is always a solution whatever the angle of propagation*

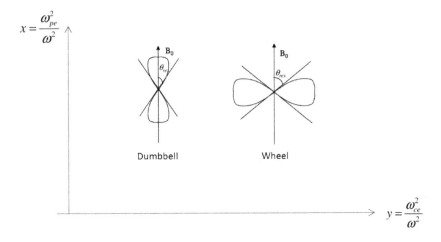

Figure 5.16. *Case where there is no solution for all the angles of propagation. On the left, the case where there is propagation only in the quasi-parallel direction, dumbbell-shaped surface. On the right, the case where there is propagation only in the quasi-perpendicular direction, wheel-shaped surface*

In the complete diagram, it can be seen that, in a given region of the plane, there can be 0, 1 or 2 solutions depending on the number of diagrams (ellipsoid, wheel, etc.) that are superimposed.

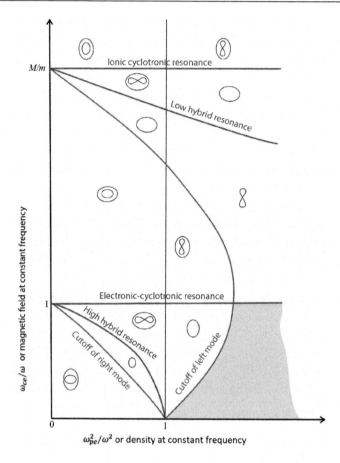

Figure 5.17. *CMA Diagram. For a color version of this figure, see www.iste.co.uk/belmont/plasma.zip*

6

Kinetic Effects: Landau Damping

Only within the limits of certain approximations, the evolution of plasma can be described with good precision by the fluid models that have been presented in previous chapters. More generally, no closed system of fluid equations can completely describe the phenomena in collisionless plasma, and the so-called "kinetic" description presented in Chapter 3 must be used. The latter describes the medium, not from the first few moments of the distribution function only, but from the full distribution function, and thus consists of not using a fluid system, but the adapted kinetic equation: the Vlasov equation. The calculation of the eigenmodes of propagation, under these conditions, is then less simple (integro-differential system); we will describe it here in detail for the basic case already presented: the Langmuir wave (or plasma wave). We will see that the kinetic effects bring a significant difference compared to the results of the fluid computations: it is the appearance of a damping of the wave called *Landau damping*.

6.1. Kinetic treatment of waves in plasmas

6.1.1. *Fluid modes and kinetic modes*

The method to calculate the eigenmodes has been presented in Chapter 5, supposing that this plasma can be correctly described by fluid equations.

Fluid descriptions involve only a few moments of the distribution function, typically density, fluid velocity and pressure. In collisionless plasmas, we know that this is *a priori* insufficient: the complete evolution of the distribution function, which is obtained thanks to the Vlasov equation, is generally necessary. This opens a much larger number of degrees of freedom than in the fluid case, since we can then initialize the function $f(v)$ in an infinite number of ways instead of only fixing the usual few fluid parameters. The important questions are: what are the linear modes of propagation in collisionless plasma described by a kinetic theory? Are some comparable to fluid modes? If so, are they exactly the same or slightly different? Are there others that can propagate?

This chapter is devoted to answering these questions. First, we will inquire about the results expected before having made the calculations. We can imagine, for example, that the kinetic modes will be much more numerous, in connection with the previous remark of the number of initialization possibilities. We may also suspect that there must remain a link between kinetic and fluid modes, at least in certain circumstances, since we know that the fluid description becomes equivalent to the kinetic description when a closure equation can be approximately justified, which occurs in some borderline cases (it can be shown, for example, that for a wave that propagates much faster than the thermal velocity of particles, the behavior of all these particles is "adiabatic").

6.1.2. *Eigenmodes: expected number and possibilities of excitation*

We have seen that the "eigenmodes" of a system are its monochromatic solutions. They are characterized by sinusoidal variations in space (k) and time (ω). The set of physical equations can be simultaneously verified for a given k only if ω satisfies a dispersion equation $D(\omega,k) = 0$. There are usually a finite number of solutions. For any non-dissipative system, that is, reversible in time, such as those that we are going to study, we seek *a priori* real solutions for ω and k. This means that, when we impose a non-evanescent solution spatially (real k), it must also be undamped in time (real ω). We have seen in Chapter 5 that typically, for a differential system consisting of N equations of the first order in time and relating N variables, the number of solutions to the problem posed is N.

If we apply this result to a kinetic system, we find that the number of solutions must be infinite, since there exists an infinity of functions $f_v(x,t) = f(v,x,t)$ to determine, each being associated with a different velocity v. This can be done through a Vlasov equation (first order in time) for each. We could also reason about the infinity of moment equations (see the hierarchy of fluid equations), which would lead to the same conclusion. We can thus guess, even before having made any calculation, that in the case of a kinetic system, the usual resolution does not directly lead to a dispersion equation in the usual sense of the term, which would have a finite number of solutions.

However, if we initiate a perturbation of the plasma without particular precautions at the microscopic level, for example, by fixing the initial perturbations of the first three moments n, $\mathbf{u}$ and $\mathbf{p}$ and supposing everywhere a Maxwellian distribution, then all the computations (analytical or numerical) and the experiments show that we obtain, in fact, a behavior close to the fluid behavior: the perturbation decomposes into a small number of eigenmodes, generally very close to the fluid modes. The main (and important) difference from the modes calculated by the fluid theory is the existence of a long-term damping of the wave. This damping even seems "universal" at asymptotic times $t \rightarrow \infty$, after a phase that depends more finely on the microscopic initialization and whose duration depends on the type of initial conditions chosen. We will explain the reasons for this damping, how it is related to initial conditions that we will describe as "generic", how to model it and thus define a "kinetic dispersion equation".

6.1.3. *Kinetic damping: a mechanical analogy*

Kinetic damping can be intuitively understood using a mechanical analogy (Belmont *et al.*, 2013). Consider a system of N coupled oscillators, consisting of weights and springs. All of these oscillators are supposed to have different eigenfrequencies when they are isolated, that is, for example, the springs are identical but all of the masses are different (let us say they are in decreasing order, to facilitate the analogy). They are coupled by the simple fact of being placed in series.

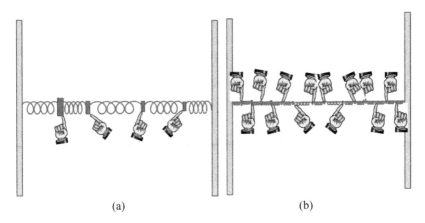

(a)   (b)

Figure 6.1. *Mechanical oscillators coupled with (a) four oscillators and (b) an increasing number of oscillators. It will be assumed that the oscillators are arranged by oscillation frequency (decreasing mass, for example). For a color version of this figure, see www.iste.co.uk/belmont/plasma.zip*

In mechanics, we very well know how to treat such a system (see section 6.8. Appendices), at least for N finite, and the difficulty will arise if we indefinitely increase the number of oscillators. The following are the main results.

If we have four oscillators in series (Figure 6.1(a): four weights and five springs), then we find that there are eight eigenmodes, that is, eight possibilities to obtain movements where the four weights oscillate with the same frequency (but with different phases). It is also found that, to excite one single eigenmode among them, it is necessary to exactly choose the initial positions of the four weights (the four initial speeds being supposed null). We thus need four fingers to initialize a single eigenmode. The four pairs of eigenmodes always have eigenfrequencies different from the four individual oscillators. If one of the individual oscillators has a frequency closer than the others to one of these eigenfrequencies, then it can be shown that its initial position must be further removed from its equilibrium position and that the oscillation of this weight will have a greater amplitude (we find a similar result in section 5.9.1, where a simpler system of two coupled oscillators is treated by different methods). If, on the contrary, the system is initially shaken in any way, then the eight eigenmodes will be simultaneously excited and the mixing of these oscillations will generally result in a somewhat disordered oscillation, with amplitude smaller than the initial amplitudes.

If we increase the number N of oscillators (Figure 6.1(b)), with $N \gg 4$, then the same kind of result remains valid. There will be $2N$ eigenmodes and, to excite only one of these eigenmodes, it will be necessary to choose the initial position of the N weights. It will therefore demand a finer initialization, with the need here to use N fingers. This obviously becomes increasingly difficult to achieve when N increases. Moreover, if a specific mode has its frequency between that of the oscillator p and that of the oscillator $p + 1$, then it will be necessary to largely displace each of the two weights p and $p + 1$ located on both sides and in opposite directions from each other.

When N goes to infinity, we understand that the exact initialization for exciting one single eigenmode becomes infinitely difficult to achieve, with a necessary number of fingers, which increases infinitely, and their fineness becomes increasingly difficult to reach. An oscillator that has the same individual frequency as one of the eigenmodes can be called a "resonant oscillator". If all the oscillator frequencies exist, then the existence or proximity of "resonant" oscillators will become inevitable regardless of the frequencies of the eigenmodes.

If we therefore do not have enough fingers, or fingers that are not fine enough, to rigorously initialize a single eigenmode, then we will necessarily excite a small packet of eigenmodes involving several surrounding oscillators. And as for any wave packet, sooner or later there will be phase mixing and therefore damping.

This is a very similar phenomenon that occurs in collisionless plasma. The kinetic damping is similarly derived from the phase mixing between infinitely close eigenmodes, which form a continuum and cannot be separated in a realistic initialization. In mechanical analogy, it was pointed out that the finesse necessary to initialize only a single eigenmode was becoming increasingly critical near the oscillator that has the same individual frequency as the eigenmode ("resonant oscillator"). The linear calculation even indicates that we should initially impose deviations that tend towards infinity and in opposite directions on both sides when we approach this resonant oscillator. We will see in the kinetic calculation that the same goes for the physics of Landau damping: the "resonant" particles play a very important role, and the impossibility of exactly initializing the distribution function near this singular speed forbids the excitation of a single eigenmode, which is at the origin of the damping.

6.2. Example of Langmuir mode

The calculations already presented in the previous chapters have shown a great variety of "fluid" eigenmodes. We find one or the other according to the hypotheses considered in each particular case: immobile ions or not, negligible thermal effects or not, electrostatic or electromagnetic perturbations, negligible ion inertia or not and so on. Kinetic calculations can naturally be made in all of these particular cases. Each time, they lead to the same conclusion that there is kinetic damping. We will present here the simplest case, which is that of the Langmuir mode, also called the "plasma mode". We thus place ourselves in the same theoretical framework as that used in Chapters 1 and 5, that is, we make the following assumptions:

– the plasma is globally neutral, $n_{0e} = Z\, n_{0i}$;

– there is no static magnetic field in the medium;

– the frequencies are high enough that the ions can be considered as immobile (we will not write in these calculations the indices e to specify that we are discussing the electrons);

– the waves are purely electrostatic;

– we suppose the monodimensional problem in direction x (we will simply note $E_x = E$ and $v_x = v$).

6.2.1. *Resonant particles*

As we have seen, thanks to the mechanical analogy above, certain particles play a particular role in the wave–particle interaction and the resulting kinetic damping: they are the "resonant" and "quasi-resonant" particles, that is, those which have a velocity equal or close to the resonance velocity, which is here $v_r = v_\varphi = \omega/k$. These particles "accompany" the electrostatic wave in its propagation. Their role will be more clearly shown in the calculation that will follow, but we can immediately formulate a correct intuition. If we indeed imagine a particle of velocity v_0 in a sinusoidal electric field of small amplitude propagating with the velocity $v_\varphi = \omega/k$, then we can understand that its trajectory $x(t)$ will be modified very differently depending on whether it has a velocity close or not to v_φ. Figure 6.2 shows some examples of such trajectories in the form $v(x)$. The trajectories $v(t)$ and $x(t)$ can be deduced from this.

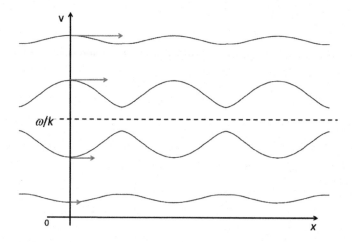

Figure 6.2. *Four examples of particle trajectories. We see that they are increasingly perturbed when the initial velocity is closer to the resonance velocity ω/k and that these perturbations are of opposite sign on either side of this velocity. For a color version of this figure, see www.iste.co.uk/belmont/plasma.zip*

If it goes much faster than v_φ, then the particle sees during its movement a field that oscillates very rapidly, and its trajectory is only slightly modulated[1], with only small accelerations and decelerations around v_0. These small variations can easily be integrated into a linear calculation.

---

1 The small perturbation of electron velocity is maximum at the maximums of the electric potential, those which have a velocity higher than v_φ being slowed down at these points and those which have a lower speed being accelerated.

The same is true if the particle has a speed much smaller than v_φ or in the opposite direction.

On the contrary, if the particle has a velocity v_0 close to v_φ, it sees a field that varies only very slowly, which gives it time to modify its trajectory much more importantly. In the limiting situation, where v_0 tends to v_φ, it would see a constant field and a constant acceleration which, in the linear calculation, would even give a velocity that would tend towards infinity. In reality, this velocity is limited by nonlinear effects. Particles whose velocity "crosses" the resonance velocity (from $v > v_\varphi$ to $v < v_\varphi$ or the opposite) have trajectories of a very different nature from linear trajectories (we will see why they can be called "trapped" particles).

These trajectories will be studied in more detail in section 6.4.2 (see Figure 6.8), but some important conclusions can be drawn right now. For a small-amplitude electric field, a linear description of the trajectories is justified for all particles that have velocities sufficiently far from the resonance velocity. However, such a description always becomes invalid in the vicinity of this velocity, where the perturbation is the most important and changes sign, as $1/(v_0 - v_\varphi)$. If we want to calculate the linear behavior of a kinetic Langmuir wave, then we will thus only be able to use the linear computation of the trajectories for the particles outside this zone of "quasi-resonant" velocity. For the particles belonging to this zone, it will be seen that we can abstain from calculating their precise trajectories and that we only have to take into account their existence via their density. This quasi-resonant zone is of very small width $|v_0 - v_\varphi| = \delta v$ when the amplitude of the electrostatic wave is very small but we cannot, despite this, ignore it since it consists of the most perturbed trajectories. The effect of these quasi-resonant particles on the wave generally finely depends on their distribution function. However, if we impose an initial "generic" disturbance, that is, without abrupt variation on the scale of this zone, then we will find the previously described "universal" kinetic damping.

6.2.2. Simulation of the Langmuir mode

Figure 6.3 shows the electrostatic energy of a Langmuir wave. To obtain it, the Langmuir wave was introduced as an initialization of a kinetic simulation[2]. The wavelength is chosen so that the predicted damping is low. The initial distributions of ions and electrons are assumed to be Maxwellian everywhere. In accordance with the kinetic theory that will be presented, we note that the amplitude of the field of the excited wave is not constant, but slowly decreases over time, with an exponential

---

2 The chosen simulation is more precisely of the "Delta_f" type, which makes it possible to have an extremely low digital noise. This allows us to study low-amplitude waves that have time to evolve long before the nonlinear effects become significant.

decay in $e^{-\gamma t}$, where $1/\gamma$ is the typical time of damping of the wave. This can also be translated by saying that the solution remains a complex exponential, but with a pulsation which is no longer real but complex: $\omega = \omega_r + i\omega_i$, with $\gamma = -\omega_i$. With these notations, the theoretical result that will be established (and to which the result of the simulation can be compared) is:

$$\frac{\omega_r}{\omega_p} \approx 1$$

$$\frac{\omega_i}{\omega_p} \approx \frac{\pi}{2} \frac{\omega_p^2}{k^2} \frac{f_0'\left(\frac{\omega_p}{k}\right)}{n_0} = -\sqrt{\frac{\pi}{8}} \frac{e^{-\frac{1}{2k^2\lambda_D^2}}}{k^3\lambda_D^3}$$

The real part of the frequency oscillates well with the known pulsation of the Langmuir wave, the plasma frequency of the electrons. In addition, we can see in these formulas an important character of "kinetic modes": the damping (imaginary part ω_i) is proportional to the derivative of the distribution function for the velocity $v = \omega_p/k$, which is the wave's phase velocity.

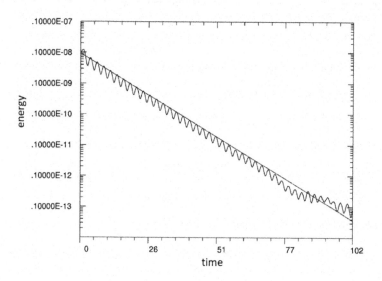

Figure 6.3. *Electrostatic energy in a kinetic simulation of a Langmuir wave. We see that this energy decreases exponentially over time (the ordinate is logarithmic)*

In the simulation, the initial perturbations of the electric field and the density are sinusoids of the given wavelength $\lambda = 2\pi/k$. Over time, this wave propagates in the simulation box with a velocity, as the calculation predicts, $v_\varphi = \omega_p/k$. It is also verified that its amplitude is not constant. This slowly decreases over time, with an exponential decay in $e^{-\gamma t}$, the rate $\gamma = -\omega_i$ being perfectly consistent, too, with the theoretical value given above for the Landau effect. Figure 6.3 shows the electrostatic energy contained in the box, which decreases at the rate 2γ (because the energy is proportional to the square of the amplitude): the line indicates the expected theoretical slope. The small parasitic oscillations only come from an imperfection of the initialization: this one did not perfectly select the wave propagating towards the right, but there is also a small percentage of the initial disturbance that propagates towards the left.

6.3. Kinetic calculation of the Langmuir mode: eigenmodes of the Vlasov/Gauss system

We will now demonstrate the previously presented results.

6.3.1. *System of equations*

The system of equations to be solved is in accordance with the physical hypotheses previously given for the Langmuir wave. It is called the Vlasov–Gauss system since the Vlasov equation, in this case, needs only the Gauss equation to form a closed system and thus to completely model the medium in the considered hypotheses. It is then a differential system with respect to the variables x and t. The functions to be determined are the electric field $E(x, t)$, which is a simple scalar in the electrostatic 1D case that we consider, and the distribution function $f(v, x, t)$, which we will note here $f_v(x, t)$ to emphasize that, from the view point of variables x and t, there exists an infinity of such functions, one for each value of the velocity v.

The equations to be solved are the Vlasov equation (i.e. a differential equation in x and t for each function f_v) and the Maxwell–Gauss equation:

$$\begin{cases} \dots \\ \partial_t f_v + v \partial_x f_v - \dfrac{eE}{m} f'_v = 0 \\ \dots \\ \partial_x E = -\dfrac{e}{\varepsilon_o}(n - Zn_i), \text{ with } n = \displaystyle\int dv \; f_v \end{cases}$$

where f'_v is the derivative of f_v with respect to v. The ellipsis are present to visually remind us that the Vlasov equation is in fact broken down into an infinitely of equations, first order in time, each dealing with a different function $f_v(x,t)$, that is, corresponding to a different velocity v. This infinity is equivalent to the infinity of fluid equations for macroscopic moments u, p, q, etc.

The first moment of the electron distribution function $n(x,t) = \int dv\ f_v(x,t)$ explicitly appears in the Maxwell–Gauss equation. This "integro-differential" aspect of the system brings some interesting properties, as well as some mathematical difficulties: indeed, this sum on the velocities couples all the Vlasov equations concerning the various functions f_v.

The ions appear in the above equations only by their density n_i. We will simply assume that this density is constant, that is, independent of x and t: $n_i(x,t) = n_{i0}$. This simple hypothesis of negligible perturbation for ions can be justified *a posteriori* because the frequencies found will be much greater than the reaction times of the ions.

We will now apply the calculation program of the classical method previously presented for fluid systems (simply differential) to calculate the eigenmodes of this kinetic system.

6.3.2. *Algebraized system*

The method first consists of linearizing this system, and this step presents no difficulty.

We have here only two types of unknowns, $E(x,t)$ and the set of $f_v(x,t)$, which are supposed to be defined by a linear equilibrium and a perturbation: $f_v = f_{v0} + f_{v1}$ and $E = E_0 + E_1$. The equilibrium state is characterized by $E_0 = 0$ and $n_0 = \int dv f_0(v) = Z n_{0i} = $ cste. The linear perturbation is characterized by $|f_{v1}| \ll f_{v0}$ and must, therefore, satisfy the following system:

$$
\begin{cases}
\ldots \\
(\partial_t + v\partial_x) f_{v1} = \dfrac{e}{m} f'_{v0}\, E_1 \\
\ldots \\
\partial_x (E_1) = -\dfrac{e}{\varepsilon_o} n_1, \text{ with } n_1 = \int dv\ f_{v1}
\end{cases}
$$

The next step is to write that all linear variations vary as $e^{ikx-i\omega t}$. We thus obtain the following algebraic system:

$$\begin{cases} \dots \\ (v - \omega / k) f_{v1} = \dfrac{e}{ikm} f'_{v0} E_1 \\ \dots \\ E_1 = -\dfrac{e}{ik\varepsilon_o} n_1, \text{ with } n_1 = \int dv \ f_{v1} \end{cases}$$

Note that these equations are based on the assumption that the particular solutions sought are "completely" monochromatic, even at the microscopic level: the distribution function is here supposed to vary in $e^{ikx-i\omega t}$ with the same frequency and wave number for all speeds v. This is a much stronger assumption than to only assume that the first few moments, n, u, p, etc., are monochromatic, as is done in fluid theory. This point will prove to be a critical point afterwards.

6.3.3. Resolution

To solve E_1, it is necessary to eliminate all the f_{v1}, that is, to express them according to E_1. A naive inversion of the Vlasov equation for f_{v1} would give:

$$f_{v1} = -\frac{ie}{km} \frac{1}{v - \omega / k} f'_{v0} E_1$$

However, we quickly see that this simple form leads to a serious difficulty: the above function is not defined for the velocity $v_r = \omega/k$, which is called the "resonance velocity". For values of v close to v_r, the function $\dfrac{1}{v - \omega / k}$ diverges, with a discontinuity in $v = v_r$, which changes it from $-\infty$ to $+\infty$. The integral in v, which we have to calculate to introduce these values of f_{v1} in the Maxwell–Gauss equation, is, for this reason, indefinite.

To better understand why this integral, supposed to give the density, is actually indefinite, we can try to calculate it as follows: before integrating, we cancel the function in a small neighborhood around the value $v = v_r$ that poses a problem, say between $v_r - \varepsilon_1$ and $v_r + \varepsilon_2$. The corresponding integral is well-defined and it depends on ε_1 and ε_2. We can then take the limit of the result when ε_1 and ε_2 tend to zero. If we choose $\varepsilon_2 = \varepsilon_1 = \varepsilon$, that is, if we symmetrically cut around v_r, then we find a limit that is effectively independent of ε and that we call the "principal value" of the integral. However, this remains a special case: if we chose, for example, $\varepsilon_2 = 2\varepsilon_1$, then we would find a different result. Hence, the integral is called "indefinite". The general solution is given by the theory of distributions:

$$f_{v1} = -\frac{ie}{km}\left[PV\left(\frac{1}{v - \omega/k}f'_{v0}\right) + \alpha n_0\, \delta(v - \omega/k)\right]\, E_1$$

The perturbation thus appears as the sum of two distributions (Figure 6.4).

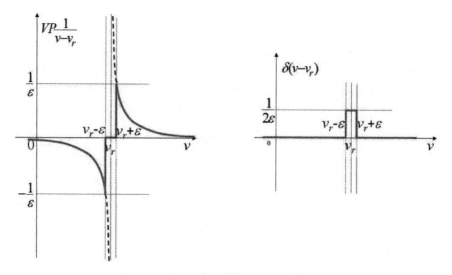

Figure 6.4. *The two functions (in red) have finite and well-defined integrals regardless of the value of ε. Their limits, when ε tends to zero, make it possible to define the so-called "principal value" and "Dirac" distributions. For a color version of this figure, see www.iste.co.uk/belmont/plasma.zip*

The "principal value" distribution $PV\left(\dfrac{1}{v - \omega/k}f'_{v0}\right)$ that appears in the first term is defined as above: this is the limit when ε tends to zero of a function equal, almost everywhere, to $\dfrac{1}{v - \omega/k}f'_{v0}$ but equal to zero in the small range of velocities between $\omega/k - \varepsilon$ and $\omega/k + \varepsilon$.

The second term involves a Dirac distribution, which can be defined as the limit when ε tends to zero of a null function everywhere except in a small velocity range between $\omega/k - \varepsilon$ and $\omega/k + \varepsilon$, where it is $\dfrac{1}{2\varepsilon}$ (integral equal to 1).

The integral of the first term is the principal value of the integral of $\dfrac{1}{v - \omega/k} f'_{v0}$. That of the second term is equal to $\alpha_1 n_0$, which is an arbitrary number. The existence of this arbitrary number reflects, in a mathematically correct way, the fact that the integral of $\dfrac{1}{v - \omega/k} f'_{v0}$ is indefinite, that is, it can take an infinite number of values: the system to be solved is satisfied regardless of the value of α_1 (this value will be determined in practice by the initial condition). We can also directly check on the Vlasov equation before inversion that, if we know a solution f_{v1}, for example, with $\alpha = 0$, then the function f'_{v1} obtained with $\alpha' \neq \alpha$ is also a solution: multiplying by $v - \omega/k$ a distribution which is localized in $v = \omega/k$ obviously does not bring any new contribution.

At this point, now that the concepts of principal value and Dirac distribution have been introduced, the delicate part of the calculation has passed. The sequel is just a series of algebraic manipulations without major difficulty. Before continuing this calculation, let us return to the calculations that have just been made and their interpretation.

NOTES.–

– The existence of the arbitrary number α must absolutely not be forgotten: it is directly at the origin of the infinite number of modes which we must find.

– The existence of two terms in the solution (principal value and Dirac value) can be easily understood. If, for a given k, we want to directly solve the linearized Vlasov equation in its differential form (i.e. before imposing the form in $e^{ikx - i\omega t}$), then we see that it implies solving a linear equation of the first order in time for the function f_{v1}. It includes a second member of "forcing" by the electric field E_1, which we will assume to be monochromatic, that is, varying in time as $e^{-i\omega t}$. It is well-known that the general solution of the complete equation in f_{v1} is the sum of a particular solution of the equation with its second member ("forced solution") and of the general solution of the equation without the second member ("free solution"). Therefore, we find that the result is the sum of two terms as we have just seen. As a particular solution, the monochromatic solution can be chosen at the same frequency ω as the forcing field. This corresponds to the first term of the previous result. The general solution of the equation without a second member, for its part, is of the form

$F(x - vt) = a\, e^{i(kx - kvt)}$. This is the ballistic transport of the initial disturbance. We see that its dependency in time is e^{-ikvt}. When, as we have done, we search for an eigenmode whose disturbances all have the same pulsation, regardless of v, this term must be taken to be zero as long as kv is not equal to ω, that is, say for $v \neq \omega/k$. However, for the velocity $v = \omega/k$, that is, when the time dependences of the free solution and forced solution are identical (which is the definition of resonance), the introduction of this second term is necessary. This is the origin of the Dirac distribution, since this occurs only for the single velocity $v = \omega/k$.

– The use of distributions, introduced here as limits of functions, seems unusual for the search of eigenmodes. In the usual mechanical problems, in particular for the fluid systems mentioned in the previous, the resolution is made without recourse to these theories. What is the specificity of the Vlasov–Gauss system that makes it necessary here? It is simply the continuous character of the variable v: there exists, for this reason, an infinite number of variables and equations and, for a given k, there exists an infinite number of resonance frequencies $\omega = kv$ (one by value of v). This means that, regardless of the eigenmode of pulsation ω, there are always particles that have the velocity v resonating with it. This is not the case with conventional discrete systems.

By inserting the f_{v1}, which we have just calculated in the Maxwell–Gauss equation, we obtain:

$$E_1 = \frac{\omega_p^2}{k^2}\left[PV \int dv\; \frac{f'_{v0}/n_0}{v - \omega/k} + \alpha \right] E_1$$

Here, the plasma pulsation defined by $\omega_p^2 = \dfrac{ne^2}{m\varepsilon_0}$ is introduced. We can rewrite this result as:

$$\left\{ 1 - \frac{\omega_p^2}{k^2}\left[Y'\left(\frac{\omega}{k}\right) + \alpha \right] \right\} E_1 = 0$$

where, to simplify the writing, we have also introduced the function Y defined by:

$$Y\left(\frac{\omega}{k}\right) = PV \int dv\; \frac{f_{v0}/n_0}{v - \omega/k} \quad \text{and therefore} \quad Y'\left(\frac{\omega}{k}\right) = PV \int dv\; \frac{f'_{v0}/n_0}{v - \omega/k}$$

In the case of a Maxwellian f_{v0} distribution, the function Y above is directly related to the "function of Fried and Conte", well-known to plasma physicists and whose real part is defined, for real ξ, by:

$$Z_r\left(\xi\right) = PV \int dx \ \frac{e^{-x^2} / \sqrt{\pi}}{x - \xi}$$

Since we are not interested in the trivial solution $E_1 = 0$, the solutions of the system must satisfy:

$$1 - \frac{\omega_p^2}{k^2}\left[Y'\left(\frac{\omega}{k}\right) + \alpha\right] = 0 \ .$$

This is the equivalent of the usual dispersion equation in eigenmode calculations. The big difference comes from the presence of the arbitrary number α, which highlights the existence of an infinite number of modes. For any ω and k, we can always find a value of α, which verifies that equation. The form of the perturbation of the distribution function is found by inserting the value of α that derives from the above equation in the expression of f_{v1}:

$$f_{v1} = \left[\frac{\omega_p^2}{k^2} PV\left(\frac{f'_{v0} / n_0}{v - \omega / k}\right) + \left(1 - \frac{\omega_p^2}{k^2}Y'\left(\frac{\omega}{k}\right)\right)\delta(v - \omega / k)\right] n_1$$

6.3.4. Concept of kinetic eigenmodes

The real ω/k solutions are not physically observable.

The previous calculation is mathematically complete, and it gives the full set of real solutions to the given linear system. As we have just seen, there is a dense infinity of such solutions. From a physical viewpoint, unfortunately, the work is not finished because none of these solutions can be observed alone, which limits their interest. We will therefore have to look for how to superimpose these solutions (make "wave packets") to make a new set of "physically observable" eigenmodes. We will see that this is possible but that the number of solutions remains infinite: we change only a base of eigenmodes for another base on which it will be more convenient to decompose the evolution of the physical system, for example, in a problem with given initial conditions.

The "non-physical" properties of the preceding solutions essentially come from the existence, in the expression of the perturbation f_{v1}, of two distributions: a main value and a Dirac value. To show this, let us reason about the functions shown in Figure 6.4, which tend towards these distributions when ε tends to zero.

– These two functions take very large values at the resonance velocity $v = \omega/k$ and in its near vicinity (changing sign for the principal value). It is the same for the perturbation f_{v1}, which is built from them. This is contrary to the physical hypotheses which led to the writing of the linear system solved since we assumed $|f_{v1}| \ll f_{v0}$. These modes even imply negative values of the distribution function $f_v = f_{v0} + f_{v1}$ for large negative values of f_{v1}, which is, of course, physically impossible.

– These two functions are extremely discontinuous: when ε tends to zero, they imply jumps of f_{v1} that tend towards infinity between two velocities distant from ε on both sides of the resonance speed.

To excite this mode alone (and not its neighbors), it would be necessary to know how to prepare the system in this way that is infinitely precise and violently discontinuous, which is infinitely difficult to achieve in practice. This means that, if we succeed in creating an initialization having the theoretical form, but where the extreme peaks are slightly blunted, then we necessarily excite not only this mode alone, but a small wave packet around it. The notion of a "wave packet" usually implies a "decay when t goes to infinity" via phase mixing. This provides a valuable clue as to how to make more natural eigenmodes.

Damped modes.

We will now look for eigenmodes such that the macroscopic quantities n_1 and E_1 are not monochromatic with real ω/k as previously, which was supposing a sinusoidal variation of constant amplitude from $-\infty$ to $+\infty$, but include a "damping", that is, an amplitude that exponentially decreases over time. Locally, we can always write this variation as $e^{-i\tilde{\omega}t}$, but with a complex pulsation: $\tilde{\omega} = \omega_r + i\omega_i$ ($\omega_i < 0$). However, we must limit this form to $t > 0$ because that would otherwise suppose a variation that tends to infinity when t tends to $-\infty$. Wave packets are thus assumed to have the following form for macroscopic quantities (Figure 6.5).

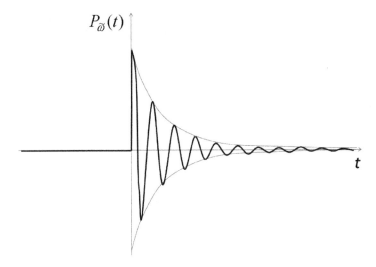

Figure 6.5. *Damped solutions: waveform shape chosen for macroscopic quantities*

This wave packet can be obtained by superimposing real eigenmodes computed previously. Since this is a continuous series of modes and not discrete modes, the sum that corresponds to this superposition is naturally an integral and not a discrete sum. The coefficients of this superposition correspond to the Laplace transform (see section 6.8.2 in this chapter's Appendices), and are easily found in tables. They are given by:

$$P_{\tilde{\omega}}(t) = -\frac{1}{2i\pi} \int d\omega \frac{1}{\omega - \tilde{\omega}} e^{-i\omega t}$$

If we apply this same superposition to f_{v1}, then we find the distribution function perturbation corresponding to these damped modes. We can show that this one is, for $t > 0$:

$$f_{v1}(t) = \frac{n_1}{v - \tilde{\omega}/k} \left[\frac{\omega_p^2}{k^2} \left(e^{-i\tilde{\omega}t} - \frac{1}{2} e^{-ikvt} \right) f'_{v0}/n_0 - \frac{1}{2i\pi} \left(1 - \frac{\omega_p^2}{k^2} Y'(v) \right) e^{-ikvt} \right]$$

To demonstrate this, we must transform the product of fractions $\frac{1}{\omega-\tilde{\omega}}\frac{1}{\omega-kv}$, which appears in the integrant into a sum, and calculate the two corresponding integrals. The integral of the first function gives without problem $e^{-i\tilde{\omega}t}$ (no pole since $\tilde{\omega}$ is not real: the main value is equal to the ordinary integral), but the integral of the second is a little more delicate since it has a pole and we have to know how to calculate its principal value. This calculation is done by passing in the complex plane for ω and integrating in this complex plane. We will not do it here, but the result is this: the main value is just half the value e^{-ikvt} that we would have if kv had a negative imaginary part (like $\tilde{\omega}$).

Two important results can be drawn from the previous calculations:

– nothing constrains the number of damped solutions as we have just defined them. These "physical" solutions are, therefore, in infinite number like the previous real solutions and as originally predicted. We even have the choice of the real part and the imaginary part of $\tilde{\omega}$, not only of its real part;

– the f_{v1} perturbation of these solutions is not monochromatic, unlike the macroscopic quantities. It includes not only a term in $e^{-i\tilde{\omega}t}$, like macroscopic quantities (as its integral n_1 in particular), but also terms in e^{-ikvt}. These correspond to the "ballistic" propagation of the initial perturbation $f_{v1}(0)$ by particles of velocity v. In the calculation of real eigenmodes, these ballistic terms had been kept only for the resonant velocity, that is, for $kv = \omega$, which preserved the monochromatic character of the solution at the microscopic level;

The perturbation f_{v1} has the factor term $\dfrac{1}{v - \tilde{\omega}/k}$. This term imposes a very particular form on the perturbation f_{v1} (a form that is independent of time). This is not a pole in v since $\tilde{\omega}$ is complex, whereas v is real. Nevertheless, this corresponds to a characteristic signature of this perturbation: in the neighborhood of $v = \omega_r/k$, its modulus increases and its argument changes by π. This signature is all the more violent than ω_i and is smaller in magnitude. The limit $\omega_i = 0$ naturally corresponds to the previous real case, which is infinitely discontinuous in this value.

We see in Figure 6.6 that, when the damping $\gamma = -\omega_i$ increases, the signature of f_{v1} around $v_r = \omega_r/k$ has a maximal modulus that decreases and a form that is increasingly blunted. However, this signature continues to exist regardless of the damping. Since this is time-independent, it means that this damped eigenmode can only be excited if this particular form is introduced from the initial condition. This supposes a very particular preparation of the plasma, which cannot be the effect of chance (possible in numerical simulation, but difficult experimentally). If we initialize the system, for example, by modulating the density and the electric field but we keep Maxwellian distributions everywhere, then it is clear that we will not be able to excite a damped mode of this kind.

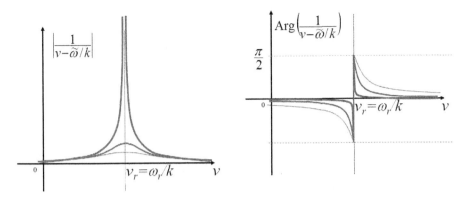

Figure 6.6. *Module and argument of the complex function* $\dfrac{1}{v - \tilde{\omega}/k}$

for different values of the imaginary part ω_i. For a color version of this figure, see www.iste.co.uk/belmont/plasma.zip

The damped Landau mode.

The above particular form does not correspond to a pole for the function f_{v1} of real v, but it nonetheless corresponds to a complex pole if we consider the analytical continuation of this function of v in the complex plane. What happens if we initialize the function f_{v1} without introducing the particular form above, that is, if we initialize by a function whose extended form does not have a complex pole (which is the general case when we do not introduce special preparations)? Among the above infinity of modes, the only ones that satisfy this condition are those that cancel the numerator (in square brackets) for the same complex value $v = \tilde{\omega}/k$, that is, such as:

$$1 - \frac{\omega_p^2}{k^2} Y'\left(\frac{\tilde{\omega}}{k}\right) - i\pi \frac{\omega_p^2}{k^2} f'_0\left(\frac{\tilde{\omega}}{k}\right) / n_0 = 0$$

In this way, we find a unique dispersion equation. This is the kinetic dispersion equation of Landau. This equation has, in fact, several solutions in the complex plane of $\tilde{\omega}$, but we are generally only interested in the least damped solution, which can be called the "Landau mode".

It should be noted that the damping thus deduced from this equation is not the consequence of any dissipation mechanism. This is the consequence of a limited choice of "regular" initial conditions, i.e. "without pole" of f_{v1} in the complex plane.

Landau damping: the limit of weak damping.

We see on the previous kinetic dispersion equation that the existence of damping ω_i comes from the presence of an imaginary term in the equation, which is proportional to the derivative $f'_0 (\tilde{\omega}/k)$. It concerns the analytical continuation of the real function f'_0 for complex values, but within the limit of weak damping $\omega_i \ll \omega_r$, we suspect that the damping will be roughly proportional to $f'_0 (\omega_r/k)$. It will then be weak, in general, for the phase velocities ω_r/k, which are large compared to the thermal velocity of the particles, where f_0 itself tends to zero. We can think, in particular, of the case of Maxwellian distribution (Figure 6.7).

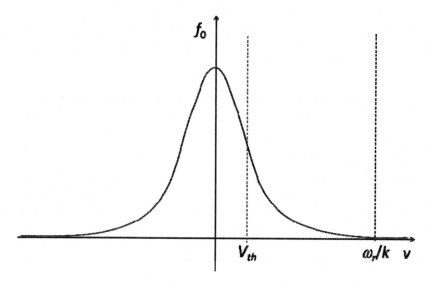

Figure 6.7. *Maxwellian distribution: for $v \gg V_{th}$, f'_0 tends to zero*

To show this, we write the dispersion equation as:

$$\varepsilon_r + i\varepsilon_i = 0$$

We obtain this by using $\varepsilon_r = 1 - \frac{\omega_p^2}{k^2} Y' \left(\frac{\tilde{\omega}}{k}\right)$ and $\varepsilon_i = -\frac{\pi \frac{\omega_p^2}{k^2} f'_0 \left(\frac{\tilde{\omega}}{k}\right)}{n_0}$ and solving in a perturbative way. If we first assume $\varepsilon_i = 0$, then the equation $\varepsilon_r = 0$ gives a first (real) solution: ω_0. The more precise solution of the dispersion equation comes at order 1 of this development: $\varepsilon_r = i\pi \frac{\omega_p^2}{k^2} f'_0 \left(\frac{\omega_0}{k}\right) /n_0$.

If we write $\omega = \omega_0 + \delta\omega$, then we have:

$$\varepsilon_r\left(\frac{\tilde{\omega}}{k}\right) \approx \varepsilon_r\left(\frac{\omega_0}{k}\right) + \frac{\delta\omega}{k}\varepsilon'_r\left(\frac{\omega_0}{k}\right) = \frac{\delta\omega}{k}\varepsilon'_r\left(\frac{\omega_0}{k}\right)$$

Thus:

$$\frac{\delta\omega}{k} \approx i\frac{\varepsilon_r}{\varepsilon'_r\left(\frac{\omega_0}{k}\right)} = i\pi\frac{\omega_p^2}{k^2}\frac{f'_0\left(\frac{\omega_0}{k}\right)/n_0}{\varepsilon'_r\left(\frac{\omega_0}{k}\right)}$$

We can still express the derivative ε'_r in the case which interests us, that is, when the majority of the particles satisfy $v \ll \omega_0/k$. In this case, the function Y can be approximated by $Y \approx -\frac{k}{\omega}$, and hence $Y' \approx \frac{k^2}{\omega^2}$ and $Y'' \approx -2\frac{k^3}{\omega^3}$. We can then approximate the denominator by $\varepsilon'_r\left(\frac{\omega_0}{k}\right) \approx -\frac{\omega_p^2}{k^2}Y''\left(\frac{\omega_0}{k}\right) = 2\frac{\omega_p^2}{\omega_0^2}\frac{k}{\omega_0}$.

Therefore:

$$\frac{\delta\omega}{k} \approx i\frac{\pi}{2}\frac{\omega_0^3}{k^3}f'_0\left(\frac{\omega_0}{k}\right)/n_0 \rightarrow \boxed{\omega_i \approx \frac{\pi}{2}\frac{\omega_0^3}{k^2}f'_0\left(\frac{\omega_0}{k}\right)/n_0}$$

We recognize in this form the different results that could be expected:

– $\delta\omega$ is purely imaginary, which means that the real part ω_r, which governs the propagation velocity ω_r/k, remains approximately that of the zero-order solution $\omega_r \approx \omega_0$ (which is the solution from $\varepsilon_r = 0$, i.e. $\sim\omega_p$), but an imaginary part ω_i also appears;

– for a positive phase velocity ω_r/k, the imaginary part ω_i is negative as soon as the derivative of the distribution function f'_0 becomes negative. It is therefore, in accordance with the calculation used to establish the formula, a damping. This is obviously the case for a Maxwellian distribution, but this is also the case, by definition, for all stable distribution functions (see section 6.4.5).

6.4. Role of resonant particles

6.4.1. *Definitions*

As has already been said, the particles whose velocity is equal to (or close to) the resonance velocity, which is here $v_r = \omega_r/k$, are called resonant (or quasi-resonant) particles. We have seen that this velocity is naturally introduced, from a mathematical viewpoint, when searching for linear eigenmodes. We have also seen that it plays a particular role in the individual trajectories of the particles. We will now take up this second point and try to more precisely understand why they play an important role in kinetic effects, and what this role is. We will see that this role can be well analyzed in two borderline cases: (1) completely nonlinear case but neglected damping; and (2) damping taken into account but with neglected nonlinear effects. We will see that the important range is defined by $\left| v - \dfrac{\omega}{k} \right| < 2\sqrt{\dfrac{e}{mk} \hat{E}}$ ($\hat{E}$ being the amplitude of the electric field oscillation) in the first case and by $\left| v - \dfrac{\omega}{k} \right| < \left| \dfrac{\omega_i}{k} \right|$ in the second case. The complete case, with damping and nonlinear effects, generally escapes any analytical treatment; it will only be mentioned quickly.

6.4.2. Nonlinear dynamics of electrons in the wave field

We can easily get an idea of the evolution of the perturbation $f_1(v)$ in the presence of a monochromatic fluctuation of the electric field. Suppose for this reason that the wave exists (not self-consistent computation) and, for simplicity, first, that its damping is negligible. In this case, it propagates without deforming at the phase velocity $v_\varphi = \omega_r/k$. We can, in this case, study the dynamics of electrons in the field of this wave in general, even without the linear assumption of the previous calculation. For this, it is convenient to calculate the particle trajectories in the wave reference frame, characterized by $x' = x - v_\varphi t$. The field is stationary in this frame, and it is written as:

$$E = \hat{E} \cos(kx' + \varphi) = \hat{E} \sin kx'$$

where $\hat{E}$ is the amplitude of the oscillation of E. Its phase φ is fixed by the origin of x'. By choosing here $\varphi = -\pi/2$, we chose to take this origin at a minimum of $|E|$ (which will also be, as we will see, a maximum of the velocity v' in this frame).

We can easily study the dynamics of an electron in such a field in all generality, that is, even without making the linear assumption that the deviation of its trajectory is small. This can be done by expressing its potential energy in this field, $U_p = -\dfrac{e}{k}\hat{E}\cos kx'$, and writing the conservation of energy during the movement:

$$\frac{1}{2}m\left(v'^2 - v'^2_0\right) = \frac{e}{k}\hat{E}\left(\cos kx' - 1\right) \rightarrow v'^2 = v'^2_0 - 4\frac{e}{mk}\hat{E}\,\sin^2\left(kx'/2\right)$$

Figure 6.8 shows the velocity v' of the particles thus determined as a function of x' for a set of initial values v'_0 given at $x' = x'_0$. We see that we can distinguish two categories of trajectories:

– far from the velocity $v' = 0$, the trajectories are "passing", that is, the velocity is slightly modulated, but remains of constant sign. The corresponding particles, therefore, continue to move to the right if they have been propelled to the right and to the left if they have been propelled to the left. In the fixed reference frame, "to the right" and "to the left" mean "faster" and "slower" than the phase velocity v_φ, respectively;

– around the velocity $v' = 0$, we have "trapping loops": the particles turn round on closed trajectories, passing periodically from $v' > 0$ to $v' < 0$.

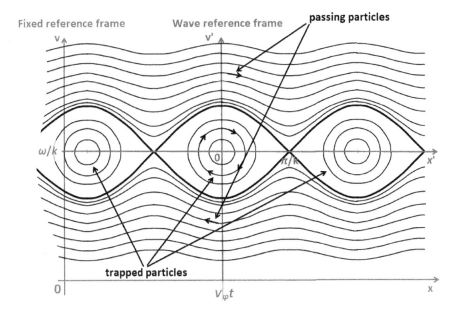

Figure 6.8. *Individual trajectories of electrons in phase space. The electric field is supposed to be monochromatic. The trajectories are calculated by imposing the conservation of energy in the wave reference frame. For a color version of this figure, see www.iste.co.uk/belmont/plasma.zip*

The trapped particles are those whose velocity in $x' = 0$ is such that $v'^2_0 < 4\dfrac{e}{mk}\hat{E}$. The v-extension of the trapping loops, therefore, depends on the amplitude of the electric field oscillation. In the linear approximation, these loops are very thin (order 1), which does not prevent them from playing an important role in the linear calculation. The average velocity of these trapped particles is equal to the phase velocity, since their instantaneous velocity oscillates around this value.

By virtue of the Vlasov equation, these trajectories are also the iso-contours of the distribution function. It is thus seen that, if the distribution function at $x' = 0$ is known, then it is known everywhere for the passing particles and, for the trapped particles, inside a trapping loop. This result can be compared with the previous linear calculation. The latter, by its linearity itself, can only determine the part sufficiently far from the resonance velocity (this linear perturbation diverges when approaching it). It therefore only takes into account passing particles, via the principal value term. For the nearest particles, trapped or not, the nonlinear effects cannot be ignored for a complete description. In the linear calculation, we do not perform this complete description, but the existence of these resonant (and quasi-resonant) particles is not ignored either. The Dirac distribution term makes it possible to introduce their density and to close the computation. We have seen that the coefficient of this important term is arbitrary in general. It is understood, in view of the nonlinear calculation above, that it suffices to change it to differently fill the trapping loops in the initial condition.

If we want to extend the nonlinear computation above to the damped modes, then we cannot make the hypothesis that the wave is a stationary structure that moves without being deformed with phase velocity v_φ. There are still trapping loops, but the decrease in the wave's amplitude causes an increasing number of particles to become "detrapped".

6.4.3. *Estimation of energy exchanges between field and particles*

Knowing the exchanges that take place between electrostatic energy and particle kinetic energy can help to better understand the physics involved in Landau damping and the specific role of resonant particles. Let us return to linear calculation. There are many ways to conduct these energy calculations. Some of these methods are Eulerian and estimate the power exchanged using the macroscopic quantities at a given point (like current j_1). Others are Lagrangian, that is, they first estimate the power received by an individual particle by following its trajectory, before averaging and summing over all the particles. Each of them highlights a different aspect of the physics involved, but none is free from technical difficulties: all of these calculations must always be handled with care (and, in particular, with great rigor in the use of Taylor expansions). We will briefly introduce one of them for the linear case. We will give the main steps of the calculation, its main results and its limits (performing the complete calculation can be an interesting but delicate exercise; the reader can do without it at first reading).

The first step is to calculate the power received by a particle moving in the imposed electric field of the damped wave. This calculation can be analytically made, provided that it introduces a Taylor expansion, assuming that the particle sees its velocity v slightly disturbed by the electric field (low amplitude $\hat{E}_1$ of the field and particle passing far enough from the resonance velocity). This assumption is consistent with the linear assumption of the previous eigenmodes computation.

We choose $E = \hat{E}_1 C$ and $C = e^{\omega_i t} cos(\omega_r t - kx)$. This is the amplitude $\hat{E}_1$, which is chosen as the small parameter of the expansion. The first orders of the expansion are thus calculated at the position $x_i(t)$ of the particle at time t, as well as its velocity v, field E and function C. The position x_{00} and the velocity x_{00} of the particle at time $t = 0$ are given. The first zero index indicates that they are zero-order quantities (zero electric field); the second zero indicates that it is the value in $t = 0$ (let us not forget that the zero order corresponds to a velocity $v_0 = v_{00} = cst$, and to an abscissa $x_0 = x_{00} + v_{00}t$, which increases with time). We can calculate the power provided by the field to the particle: $P = qEv$. The complete result is rather complicated, but most of the terms disappear if we integrate on all initial positions x_{00} of the particles with the same initial velocity v_{00}. The zero order P_0 is trivially zero. The order one (P_1) is not zero for a given value of the initial position x_{00}, but its mean value over x_{00} is zero if we assume that the particles are initially homogeneously distributed in space. The term of order two (P_2) is non-zero, and its average value is non-zero, too.

Considering, moreover, the average value over a pseudo-period $T = 2\pi/\omega_r$, we can put the result thus obtained as follows:

$$\langle P_2 \rangle \approx -\omega_i \frac{\varepsilon_0 \hat{E}_1^2}{2n_0} \frac{\omega_p^2}{k^2} \frac{v_{00}^2 - \omega_r^2 / k^2 - \omega_i^2 / k^2}{\left[(v_{00} - \omega_r / k)^2 + \omega_i^2 / k^2 \right]^2}$$

Let us begin by considering this result for particles sufficiently far from the resonance velocity so that we can neglect $\left| \frac{\omega_i}{k} \right|$ with regard to $\left| v_{00} - \frac{\omega_r}{k} \right|$. We will call these particles "non-resonant". The formula is then simplified to:

$$\langle P_{2NR} \rangle \approx -\omega_i \frac{\varepsilon_0 \hat{E}_1^2}{2n_0} \frac{\omega_p^2}{k^2} \frac{v_{00}^2 - \omega_r^2 / k^2}{(v_{00} - \omega_r / k)^4}$$

In the formula thus simplified, we find a non-trivial property of these particles: for $\omega_i < 0$ (damping), all those which have a velocity v_{00} greater, in magnitude, than the resonance velocity gain, on average, energy. Conversely, those with a lower velocity lose, on average, energy. For a decreasing distribution function such as a

Maxwellian, there is therefore a greater number of particles that lose energy in favor of electrostatic energy than particles that gain. It can be seen that these non-resonant particles alone cannot explain why the electrostatic energy decreases.

Closer to the resonance velocity, the formula shows that a more complex transition involving resonant and quasi-resonant particles exists. It is logical that this small neighborhood plays an important role because, as we see on the simplified formula, the energy exchange increases considerably (it would diverge if we kept this simplified formula until ω/k) with a sign different on both sides of the resonance velocity. This behavior is very reminiscent of the perturbation of the distribution function that we encountered when calculating eigenmodes.

The second step is to estimate the power provided by the electric field to all particles, regardless of their initial velocity v_{00}. This requires introducing the distribution function $f(v_{00})$ of the initial velocities and thus formalizing the previous intuitive ideas about the numbers of particles that gain and lose energy. We will see that locally, in the vicinity of the resonance velocity, this naturally introduces the slope $f'(\omega/k)$ of the distribution function.

It is therefore necessary to estimate the integral:

$$\langle P_{tot} \rangle = \int dv_{00} \langle P\, f(v_{00}) \rangle$$

That is, at order 2:

$$\langle P_{2tot} \rangle = \int dv_{00} \langle P_2 \rangle f_0(v_{00}) + \int dv_{00} \langle P_1 f_1(v_{00}) \rangle$$

The second integral is non-zero only if the initial perturbation f_1 of the distribution is spatially correlated, for each velocity v_{00}, with the power fluctuation P_1. This corresponds to a very particular preparation of the plasma in this initial condition, and we have seen that it is precisely this infinity of choice in the initial microscopic condition that makes it possible to find an infinite number of eigenmodes, without privileged frequency or privileged damping. The "standard" case, which leads to classical Landau damping, is the one where such a correlation does not exist and where this integral is therefore zero. For any initial condition, even if the integral is not zero at the first moments following the initial condition, it generally becomes asymptotically null when t tends to infinity, by phase mixing (see the notion of a wave packet).

To estimate the first term, we have to compute an integral, which presents difficulties of the same order as that of the computation of the eigenmodes, the non-resonant region corresponding to the principal value, and the "resonant" (and quasi-resonant) part corresponding to the contribution of the Dirac distribution. In the implicit hypotheses $\omega_i \ll \omega_r$, the result is:

$$\langle P_{2tot} \rangle = -\pi \frac{\omega_p^3}{k^2} \frac{\varepsilon_0 |E|^2}{2} f_0' \left(\frac{\omega}{k} \right) / n_0$$

Knowing that the total power received by the electrons is equal to the power lost by the electric field, we can deduce:

$$-\pi \frac{\omega_p^3}{k^2} \frac{\varepsilon_0 |E|^2}{2} f_0' \left(\frac{\omega_p}{k} \right) / n_0 = -2\omega_i \frac{\varepsilon_0 |E|^2}{2} \rightarrow \omega_i = \frac{\pi}{2} \frac{\omega_p^3}{k^2} f_0' \left(\frac{\omega_p}{k} \right) / n_0$$

We thus find the same expression of Landau damping as we have already established (the frequency $\omega = \omega_0$ of the eigenmode is indeed equal to the plasma frequency ω_p in the hypothesis considered $\omega_i \ll \omega_r$). The above calculation, in spite of its difficulties, can be considered as another method to establish it and it has the advantage to underline a new point of the physics brought into play. In this linear computation, the particles that initially belong to the f_1 perturbation exchange energy with the electric field and this average exchange is of order 1 during their ballistic movement at constant velocity v_{00}. However, in the Landau hypothesis of an initial condition, which ends up being forgotten by phase mixing, the average value of this exchange ends up being canceled. It then remains only to take into account the exchange of energy with the particles of the population f_0, whose motion, at the same order, is forced by the electric field. This allows us to select the Landau solution among the infinite number of solutions to the problem.

A final point deserves to be underlined regarding energy exchanges. We have seen that the non-resonant particles exchanged with the electric field a power that did not have the right sign. Is it a paradox? If we make the calculation by separating the two contributions to the integral, non-resonant particles and "resonant" (and quasi-resonant) particles, then we obtain the following interesting results:

<P_{2res}> = −2 P_{2E}

<P_{2nr}> = P_{2E}

$$\rightarrow \langle P_{2res} \rangle = -[P_{2E} + \langle P_{2nr} \rangle]$$

where P_{2E} is the total average energy received by the electric field (second order), <P_{2res}> is the energy received by the quasi-resonant particles and <P_{2nr}> is the energy received by the non-resonant particles.

Non-resonant particles effectively exchange half the resonant particles in absolute value, but in the other direction. It can be said that the resonant particles gain their energy at the expense of the electrostatic energy and the energy of the non-resonant particles, these two contributions being equal. The sum of the electrostatic energy and the kinetic energy of non-resonant particles is sometimes called "wave energy". With this vocabulary, we can say that the energy gained by the resonant particles is lost by the wave energy.

6.4.4. *Number of resonant particles tending towards zero: fluid limit*

We can guess that the role of the resonant particles decreases when they are less numerous, that is, when the phase velocity is large compared to the mean-squared velocity of the particles of the distribution (Figure 6.7). By making a Taylor expansion of the previous result for $\omega/k \gg v$, we should find again the fluid result. Let us check it. The integral to calculate is:

$$I(\omega/k) = \int dv \frac{f'_o/n_o}{v-\omega/k} = \int dv \frac{f_o/n_o}{(v-\omega/k)^2} = \left\langle \frac{1}{(v-\omega/k)^2} \right\rangle$$

where we made an integration by parts to have the second equality and used the fact that the equilibrium distribution function is null at the integration terminals.

By carrying out the indicated expansion $kv/\omega \ll 1$, we get:

$$I(\omega/k) = \frac{k^2}{\omega^2} \left\langle \frac{1}{(1-kv/\omega)^2} \right\rangle = \frac{k^2}{\omega^2} \left\langle 1+2kv/\omega+3k^2v^2/\omega^2 +... \right\rangle$$

Thanks to the Taylor expansion, we no longer have a pole at the denominator and the integral (or mean) is well-defined: in a resting medium (where the average velocity is zero), knowing that $<v^2> = V_{the}^2$, we finally have:

$$I(\omega/k) = \frac{k^2}{\omega^2} \left[1+3k^2v_{the}^2/\omega^2 +... \right]$$

which provides the dispersion equation:

$$1-\frac{\omega_{pe}^2}{k^2}I(\omega,k) = 0 \Rightarrow \omega^2 = \omega_{pe}^2 \left[1+\frac{3k^2v_{the}^2}{\omega^2} + \cdots \right] \approx \omega_{pe}^2 + 3k^2v_{the}^2$$

The first two terms of the expansion therefore make it possible to find the fluid result, with a value $\gamma = 3$ determined here in a self-consistent way (whereas it had to be posited arbitrarily in the fluid computation, via a closure equation; see Chapter 5). Naturally, this result remains incomplete since it lacks the effect of the existing pole in the integral, but we know that we have considered here a case where this effect is small.

NOTE.– In the fluid calculation, this result was found by assuming that the compressions and decompressions of the electrons were adiabatic 1D. It can be seen here that this result is valid in the limit of the large phase velocities with respect to the typical velocity of the particles, the thermal velocity. The adiabatic approximation is therefore valid for k small before ω/V_{th}. Since $\omega \sim \omega_{pe}$, this amounts to writing $k \ll 1/\lambda_{De}$ (wavelengths long with respect to the Debye length).

6.4.5. Reverse Landau effect

Previous calculations have shown that a plasma mode that has a phase velocity ω_r/k is always damped when $f'_0(\omega_r/k) < 0$. Conversely, when $f'_0(\omega_r/k) > 0$, we can show that the formula established in the damped case remains valid and that we therefore have a growth $\omega_i > 0$. This shows that all the distribution functions that have a positive slope somewhere are potentially unstable (Figure 6.9). Kinetic instabilities are essentially different from fluid instabilities in that they come from the characteristics of the distribution function (velocity space) and can, therefore, exist in a totally homogeneous medium in real space.

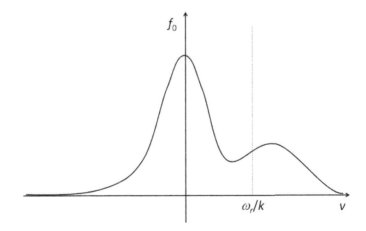

Figure 6.9. *Potentially unstable distribution function*

The excitation and evolution of this instability can be shown by numerical simulations, by initializing the plasma with a distribution function having a positive slope. Figures 6.10, 6.11 and 6.12 show the growth of unstable waves, as well as their saturation via nonlinear effects.

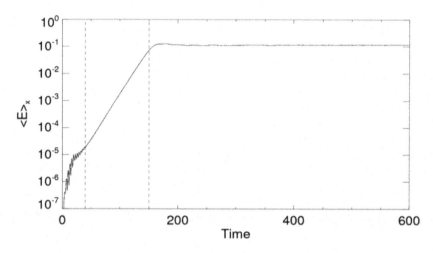

Figure 6.10. *Growth of the amplitude of the wave with time until saturation. This growth is exponential between the two dashed lines*

Figure 6.10 shows the amplitude of the electric field of the wave, which increases with time. During this growth, the slightly faster particles give up energy to the wave, the number of particles with a velocity slightly lower than the wave velocity increases and the distribution function flattens with the time around velocities close to the phase velocity (Figure 6.11). When the slope of the distribution function becomes zero (formation of a "plateau"), the wave ceases to grow and the amplitude of the field remains constant, as can be seen in Figure 6.10. This is called saturation of the wave.

In Figure 6.12, we see in more detail what happens to the electrons. It shows the velocity of electrons as a function of space at different times. Each point of the figure corresponds to the position and velocity of a set of electrons, in phase space. The particles in the bump of the distribution function are denoted by the blue line around 5 (in arbitrary units): we can see that, with time, the particles with a speed close to the phase velocity will interact with the wave and change the distribution function.

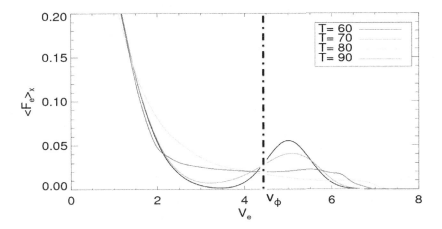

Figure 6.11. *Evolution of the distribution function over time. At the initial time, the phase velocity of the wave v_ϕ is in the zone with a positive slope, but with time, the function evolves and flattens, until the system becomes stable. For a color version of this figure, see www.iste.co.uk/belmont/plasma.zip*

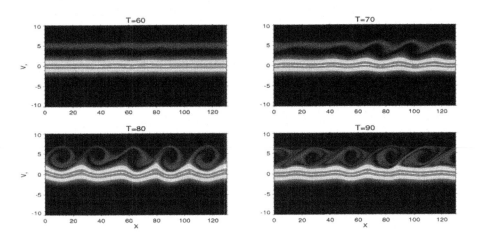

Figure 6.12. *Evolution of the phase space of electrons over time. For a color version of this figure, see www.iste.co.uk/belmont/plasma.zip*

6.5. Other methods of calculating eigenmodes

We presented above a calculation of the eigenmodes inspired by van Kampen (1967). It clearly shows that the kinetic damping originates from "phase mixing" and that it can be treated via the theory of distributions and the notion of "wave packets". We have also seen that the computation can be presented in its energetic form and that it leads to the same universal damping if we assume that the particles belonging initially to the perturbation f_1 have a contribution to energy exchanges between field and particles, which tends towards zero when the time increases ("forgotten" initial condition). Calculations of this type are found in Chen or Davidson (1972). The two types of calculation show that the complete solution depends on the initial conditions and, in particular, on the initial disturbance around the resonance velocity.

However, the classic method for calculating Landau damping is that introduced by Landau in 1946. Since it is true that the solution depends on the initial conditions, the most direct mathematical method to find this solution is to use a Laplace transformation in time, which explicitly introduces these initial conditions. We show in section 6.8.1 of this chapter's Appendices how to solve a simple differential system by the Laplace transformation. We will indicate in section 6.8.2 how this method can be applied to the Vlasov/Gauss system. This requires some mastery of integration in the complex plane. This Landau calculation was a great revolution in the physics of the time. Nevertheless, its mathematical difficulties sometimes tend to make it a little abstract and hide the physical stakes that we have just analyzed. In any case, the condition of an initial perturbation "without complex pole" for the distribution function is, indeed, present also in this calculation, which corresponds to the choice of an initial condition "without special preparation" that we have highlighted.

6.6. Other damped kinetic modes and other resonances

6.6.1. *Landau effect for modes other than the Langmuir mode*

The method that has just been presented to introduce the notion of kinetic damping has been presented in the context of the Vlasov/Gauss system, in the hypotheses that, in fluid, lead to a single mode: the Langmuir mode ($\omega = \omega_p + \delta\omega$), that is, essentially, immobile ions forming a still background (justified by the high frequencies considered). However, when one takes into account the ion motion, the preceding chapters have shown that there are many other eigenmodes of the plasma. All of these modes are subject, like the Langmuir mode, to kinetic damping.

For any kinetic mode, the principle of calculations is identical to those just presented. The common point is naturally the Vlasov equation, or the two Vlasov equations when ions and electrons have to be kinetically treated. In the electrostatic case, this gives us the perturbations f_1 of each species as a function of E_1, with the difficulties that have been described, due to the presence of resonant particles. The only thing that changes is how to express E_1 according to these two distributions (via integrals). This does not bring any change of principle. In the complete electromagnetic case, the perturbations f_1 of the two species depend on E_1 and B_1. Closing the system then requires involving Maxwell equations other than the Gauss equation alone, without any change in the principle either.

As an example of the electrostatic case, the method can be applied to the case of ion acoustic waves. Knowing the fluid solution (see Chapter 5), we know that these waves are characterized by $V_{thi} < \frac{\omega}{k} < V_{the}$. Within this limit, we obtain the dispersion equation:

$$D(\omega,k) = 1 - \frac{\omega_{pi}^2}{\omega^2}\left(1 + \frac{3k^2 v_{thi}^2}{\omega^2}\right) + \frac{1}{k^2 \lambda_{De}^2} + i\pi \frac{\omega_{pi}^2}{k^2} f'_{0,i}\left(\frac{\omega_r}{k}\right) / n_{0,i} + i\pi \frac{\omega_{pe}^2}{k^2} f'_{0,e}\left(\frac{\omega_r}{k}\right) / n_{0,e} = 0$$

To arrive at this formula, two equilibrium functions have been used for Maxwellian ions and electrons:

$$f_{0e}(v) = \frac{1}{\sqrt{2\pi v_{the}^2}} \exp\left(-\frac{v^2}{2 v_{the}^2}\right) \text{ et } f_{0i}(v) = \frac{1}{\sqrt{2\pi v_{thi}^2}} \exp\left(-\frac{v^2}{2 v_{thi}^2}\right)$$

This equation allows us to identify the eigenfrequency and the Landau damping of the ion acoustic waves:

$$1 - \frac{\omega_{pi}^2}{\omega_r^2}\left(1 + \frac{3k^2 v_{thi}^2}{\omega_r^2}\right) + \frac{1}{k^2 \lambda_{De}^2} = 0$$

$$\omega_r^2 = \frac{k^2 c_{s0}^2}{\left(k^2 \lambda_{De}^2 + 1\right)} + k^2 c_{s0}^2\left(\frac{3T_i}{T_e}\right) \text{ où } c_{s0}^2 \equiv \frac{ZT_e}{m_i}$$

$$\Rightarrow \frac{\omega_i}{\omega_r} = \frac{k^2 \lambda_{De}^2}{k^2 \lambda_{De}^2 + 1} \left(i\pi \frac{\omega_{pi}^2}{2k^2} f'_{0,i} \left(\frac{\omega_r}{k} \right) / n_{0,i} + i\pi \frac{\omega_{pe}^2}{2k^2} f'_{0,e} \left(\frac{\omega_r}{k} \right) / n_{0,e} \right)$$

$$= -\frac{\sqrt{\pi/8}}{\left(k^2 \lambda_{De}^2 + 1 \right)^{3/2}} \left[\left(\frac{ZT_e}{T_i} \right)^{3/2} \exp \left(-\frac{ZT_e/T_i}{2(1 + k^2 \lambda_{De}^2)} \right) + \sqrt{\frac{m_e}{m_i}} \right]$$

We see that the condition $\omega_i \ll \omega_r$ is satisfied only if $T_i \ll Z T_e$, which gives us a condition of existence of the ion acoustic waves. For this reason, the term $3T_i/T_e$ for the real part of the frequency has been neglected in the damping calculations of the wave. We also note that we have two terms that will contribute to the damping of ion acoustic waves: the first is due to the damping by the ions, and the second to the damping by the electrons. Each time, the derivative of the distribution function of each species, evaluated at the phase velocity of the wave, intervenes.

6.6.2. *Cyclotron resonances*

The kinetic effects we have just studied are due to a type of resonance called "Landau resonance" or "Cerenkov resonance". They correspond to the singularity that appears at the velocity $v = \omega/k$. In magnetized plasma, a more complete calculation shows that this resonance is not the only one. There is a resonance for all velocities v_n whose parallel component satisfies:

$$v_{n//} = \frac{\omega - n\omega_c}{k_{//}}$$

The Landau resonance is only the zero-order resonance of this series. The others are called "cyclotron resonances". The Landau resonance concerns an interaction between the parallel electric field and the parallel motion of the particles. The cyclotron resonances concern interactions between the perpendicular electric field and the perpendicular motion of the particles (rotation) around the magnetic field.

6.6.3. *Other resonances*

Note that other resonances may exist. Each corresponds to a possible periodic motion of the particles. Cyclotron resonances correspond to the periodic motion of gyration around the magnetic field. As discussed in Chapter 2, other periodic motions are possible when the particle moves in a variable magnetic field, for example, the bounce motion between two mirror points. The corresponding resonance, which depends on the initial velocity of the particles, is called "bounce resonance" ω_b. The waves having the frequency ω_b will make these particles play a particular role, which will generally result in some damping.

6.7. Damping and reversibility

We might think that the "universal" decay that characterizes the Landau damping at asymptotic times (i.e. when t goes to infinity) is the symptom of a certain irreversibility of the plasma during its evolution. This is not the case. We have seen that this damping derives from the Vlasov equation, which is perfectly reversible. It is, in fact, just a consequence of the selection that has been done in the initial conditions: we have seen that there are undamped (real) solutions, and also that they are rejected because they are inaccessible physically; they would require initial conditions that are mathematically singular and that cannot be realized. It is, therefore, not during the dynamic evolution that the "universality" of the decay is situated, but rather in the choice of the initial condition.

EXAMPLE.– If we place in a box, particles without interaction between them but which bounce elastically on the walls, then we obviously have a reversible system. If we place all the particles in the center of the box with random velocities in all directions, then we quickly arrive at a distribution that seems random and homogeneous. And yet, if we know how to make an initial condition that corresponds to the exact final situation of the previous experiment and if we reverse the velocity of each individual particle, transforming v into $-v$, then we will exactly return to the initial condition at the end of the same time, which is the definition of reversibility. On the contrary, if we only know how to make coarser initial conditions, for instance, if we only know how to impose the average density of the particles and their average velocity in pixels of a certain size (even small), then we will never be able to return to the initial condition and the result will appear irreversible. In this example, we see that it is the choice of "accessible" initial conditions that determines the "one-way" evolution.

In the case of Landau damping, however, it can be shown that an infinitesimal level of collisions in the plasma is enough to make the phenomenon truly irreversible (although these collisions are not the cause).

We have seen that the actual linear damping depends on the distribution function f_0 and the initial perturbation f_1. This is also true when we leave the linear domain and even when we consider potentials other than the electrostatic potential. The classes of the functions f_0 and f_1, which lead to a damping in this very general case, are extremely broad. They were the subject of advanced mathematical studies that earned Cédric Villani the Fields Medal in 2010.

6.8. Appendices

6.8.1. *Landau-Laplace method of finding eigenmodes*

In fluid problems, the equation system that governs eigenmodes always has a relatively small number of solutions. The different methods that can be used to find them are then roughly equivalent. In kinetics, the number of eigenmodes tends towards infinity and the generalizations of the various methods then present more or less difficulties. The use of the Laplace transform has proved to be one of the most effective in this case.

To illustrate the different resolution methods, we will use a very simple example of coupled oscillators for which none of the methods presented pose any technical difficulty. The results can in this case be easily obtained by each of them and compared. This example is the same as the one in Figure 6.2, but in the simplest case where there are only two oscillators, identical to each other: two masses of the same value m, three springs of the same stiffness K. Therefore, there are four eigenmodes (two pairs).

The equation of the problem is presented in Box 6.1.

The deviation of mass 1 (left) from its equilibrium position is denoted by x_1, and that of mass 2 is denoted by x_2. Mass 1 is, therefore, subjected to a force $-Kx_1$ from the first spring and a force $-K(x_1-x_2)$ from the second. We can apply the same reasoning to mass 2, and we deduce the system written below.

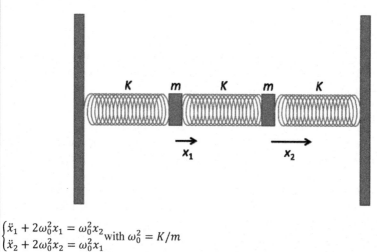

$$\begin{cases} \ddot{x}_1 + 2\omega_0^2 x_1 = \omega_0^2 x_2 \\ \ddot{x}_2 + 2\omega_0^2 x_2 = \omega_0^2 x_1 \end{cases} \text{with } \omega_0^2 = K/m$$

Box 6.1. *System of two identical coupled oscillators*

6.8.1.1. *Recall of the classical method in the general case*

Once the physical equations are written, we are dealing with a system of differential equations involving the functions of time $x_1(t)$, $x_2(t)$, etc., and their derivatives. When we look for eigenmodes, we seek solutions where all the functions sinusoidally vary with the same pulsation ω, that is, $x_i(t) = \hat{x}_i e^{-i\omega t}$. We therefore denote by $\hat{x}_i$ the amplitude of this variation. Inserting it into the system of differential equations, we obtain a system of linear and homogeneous equations for x_i, which involves ω and the parameters of the system (possibly k). We can put this linear system in the form $\mathbf{M}.\mathbf{X} = 0$, where $\mathbf{X}$ is the vector containing all the x_i and $\mathbf{M}$ is a matrix. We therefore have a non-trivial solution only if the determinant of the matrix $\mathbf{M}$ is zero. The equation $\det(\mathbf{M}) = 0$ (which contains ω) is the so-called dispersion equation if the system involves a spatial variation characterized by a wave number k. It allows us to get the N eigenmodes in the form of N values ω_n of ω with which are associated eigenvectors $\mathbf{X}_n$, which indicate the polarization, that is, the values of the functions x_i (to a common proportionality coefficient, which indicates the amplitude of the wave).

$$x_i(t) = \hat{x}_i e^{-i\omega t}$$

$$\dot{x}_i = -i\omega x_i$$

$$\ddot{x}_i = -\omega^2 x_i$$

System of equations: $\begin{pmatrix} \omega^2 - 2\omega_0^2 & \omega_0^2 \\ \omega_0^2 & \omega^2 - 2\omega_0^2 \end{pmatrix} \cdot \begin{pmatrix} x_1 \\ x_2 \end{pmatrix} = 0.$

Eigenmodes: $\omega = \pm\omega_0$, avec $x_2 = x_1$ et $\omega = \pm\sqrt{3}\omega_0$, avec $x_2 = -x_1$.

General solution: $\begin{cases} x_1(t) = a\,e^{-i\omega_0 t} + b\,e^{+i\omega_0 t} + c\,e^{-i\sqrt{3}\omega_0 t} + d\,e^{+i\sqrt{3}\omega_0 t} \\ x_2(t) = a\,e^{-i\omega_0 t} + b\,e^{+i\omega_0 t} - c\,e^{-i\sqrt{3}\omega_0 t} - d\,e^{+i\sqrt{3}\omega_0 t} \end{cases}$

Initial conditions: $x_1(0) = x_{10}$; $x_2(0) = 0$; $\dot{x}_1(0) = \dot{x}_2(0) = 0$.

➜ $a = b = c = d = x_{10}/4$

$$\begin{cases} x_1(t) = \dfrac{x_{10}}{2}\left[\cos \omega_0 t + \cos \sqrt{3}\omega_0 t\right] \\ x_2(t) = \dfrac{x_{10}}{2}\left[\cos \omega_0 t - \cos \sqrt{3}\omega_0 t\right] \end{cases}$$

Box 6.2. *Resolution by the classical method*

The general solution of the initial differential system can thus be finally written, for each function x_i:

$$x_i(t) = \sum \hat{x}_{in} e^{-i\omega_n t}$$

The constants $\hat{x}_{in}$ reflect the fact that the general solution is a linear combination of the eigenmodes of the system. These constants are determined from the particular initial conditions that are given for the problem under study. The application of this general method in the case of the two coupled oscillators is given in Box 6.2. It will be assumed here that only the first mass is displaced initially (with $x_1 = x_{10}$), and no initial velocity is given to either mass.

6.8.1.2. Fourier transformation resolution

The previous method strongly recalls the notion of Fourier transformation. The calculation can, in fact, be presented in a slightly more general way using this transformation. This is shown in Box 6.3, which, as we see, is formally very similar to Box 6.2.

Recall first of all the definition of the Fourier transformation. Denoting by $\tilde{x}(\omega)$ the Fourier transform of $x(t)$, this definition is written as:

$$\tilde{x}(\omega) = TF[x(t)] = \frac{1}{2\pi} \int_{-\infty}^{\infty} x(t) e^{i\omega t} dt$$

The inverse transformation is calculated by:

$$x(t) = TF^{-1}[\tilde{x}(\omega)] = \int_{-\infty}^{\infty} \tilde{x}(\omega) e^{-i\omega t} d\omega$$

The application of the Fourier transformation to the differential system clearly leads to the same system as before, but this time with the Fourier transforms $\tilde{x}_i(\omega)$. As before, the system being homogeneous, it yields $\tilde{x}_i(\omega) = 0$ for all the values of ω except for the four discrete values ω_n, which correspond to the previous eigenmodes and cancel the determinant. This means that the general solution is written in Fourier space: $\tilde{x}_i(\omega) = \sum_n \hat{x}_{in} \delta(\omega - \omega_n)$.

To find the eigenmodes as a function of time, it is necessary to invert the transformation:

$$x_i(t) = TF^{-1}[\tilde{x}_i(\omega)] = \int_{-\infty}^{\infty} \sum_n \hat{x}_{in} \delta(\omega - \omega_n) e^{-i\omega t} dt = \sum_n \hat{x}_{in} e^{-i\omega_n t}$$

Thus, we again find the same result as the classical method. The linear combination of the eigenmodes is introduced naturally in this computation by the fact that the general solution, in Fourier space, is a "Dirac comb", that is, the sum of four Dirac distributions.

$$x_i(t) \rightarrow \tilde{x}_i(\omega)$$

$$\dot{x}_i \rightarrow -i\omega\tilde{x}_i$$

$$\ddot{x}_i \rightarrow -\omega^2\tilde{x}_i$$

System of equations: $\begin{pmatrix} \omega^2 - 2\omega_0^2 & \omega_0^2 \\ \omega_0^2 & \omega^2 - 2\omega_0^2 \end{pmatrix} \cdot \begin{pmatrix} \tilde{x}_1 \\ \tilde{x}_2 \end{pmatrix} = 0.$

Same eigenmodes:

$\omega = \pm\omega_0$, avec $\tilde{x}_2 = \tilde{x}_1$

$\omega = \pm\sqrt{3}\omega_0$, avec $\tilde{x}_2 = -\tilde{x}_1$

with $\tilde{x}_i(\omega) = \sum_n \hat{x}_{in}\delta(\omega - \omega_n)$.

Inverse Fourier transformation:

$$x_i(t) = TF^{-1}[\tilde{x}_i(\omega)] = \int_{-\infty}^{\infty} \sum_n \hat{x}_{in}\delta(\omega - \omega_n) \, e^{-i\omega t} \, dt = \sum_n \hat{x}_{in} \, e^{-i\omega_n t}$$

Initial conditions: $x_1(0) = x_{10}; x_2(0) = 0; \dot{x}_1(0) = \dot{x}_2(0) = 0$.

➔ $\hat{x}_{in}$=x10/4

$\begin{cases} x_1(t) = \frac{x_{10}}{2}[\cos \omega_0 t + \cos \sqrt{3}\omega_0 t] \\ x_2(t) = \frac{x_{10}}{2}[\cos \omega_0 t - \cos \sqrt{3}\omega_0 t] \end{cases}$

Box 6.3. *Resolution by Fourier transformation*

6.8.1.3. *Resolution by Laplace transformation*

The definition of the Laplace transformation resembles that of the Fourier transformation:

$$\tilde{x}_i(\omega) = TL[x_i(t)] = \frac{1}{2\pi} \int_0^{\infty} x_i(t)e^{i\omega t} \, dt$$

The expression of the inverse Laplace transformation is:

$$x_i(t) = TL^{-1}[\tilde{x}_i(\omega)] = \int_C \tilde{x}_i(\omega)e^{-i\omega t}\, d\omega$$

NOTE.– This transform is more generally presented with the variable $p = -i\omega$ rather than ω, but this does not change anything important to the principle.

– We see that the direct transformation has the lower bound $t = 0$ and not $-\infty$ like the Fourier transform. It thus explicitly involves the initial conditions, which explicitly appear in the calculation of the derivatives. Indeed, by integrating by parts the derivative transform, we find:

$$\tilde{\dot{x}}_i(\omega) = \frac{1}{2\pi}\int_0^\infty \dot{x}_i(t)e^{i\omega t}\, dt = -i\omega\tilde{x}_i(\omega) - \frac{1}{2\pi}x_i(0)$$

$$\tilde{\ddot{x}}_i(\omega) = -\omega^2\tilde{x}_i(\omega) + \frac{i\omega}{2\pi}x_i(0) - \frac{1}{2\pi}\dot{x}_i(0)$$

– For the inverse transform, it is an integration in the complex plane and not on the real axis like the Fourier transform. This amounts to splitting the signal on damped modes and not on real modes. The contour C can have any form provided that it passes over all the poles of $\tilde{x}_i(\omega)$, which are again, as we shall see, the four values ω_n that cancel the determinant of the same matrix **M** as before.

By applying the Laplace transform to the original system, we obtain this time a system with a second member: $\mathbf{M}.\tilde{\mathbf{X}} = \mathbf{A}$, the vector **A** of the second member containing the initial conditions. This system resolves into $\tilde{\mathbf{X}} = \mathbf{M}^{-1}.\mathbf{A}$. It only remains to calculate the inverse transform to obtain **X** as a function of time. Thus, we find by this method the eigenpulsations found by the preceding methods: these are the poles of the inverse Laplace transformation.

The application of this Laplace method is shown in Box 6.4. It can be seen that, in this form, the result is a superposition of monochromatic solutions with complex frequencies, but that these are in infinite number (continuous integral), the different frequencies being taken along a contour C whose choice implies a certain arbitrariness. In order to perform the integral, we need to have some basic knowledge about the integration of complex variable functions, in particular, the residue method (see the Cauchy integral). The choice of contour shown in Figure 6.14 then leads to the result in a simple way if we know that each of the poles makes a contribution equal to $-2i\pi$ times the value of the corresponding residue corresponding to this pole (the minus sign comes in because of the rotation in the retrograde direction around the pole). The contribution of all the horizontal elements at the bottom of the contour (ω_i tending towards $-\infty$) is null, like that of the vertical parts, which cancel themselves two by two.

$$x_i(t) \rightarrow \tilde{x}_i(\omega)$$

$$\dot{x}_i \rightarrow -i\omega \tilde{x}_i(\omega) - \frac{1}{2\pi} x_i(0)$$

$$\ddot{x}_i \rightarrow -\omega^2 \tilde{x}_i(\omega) + \frac{i\omega}{2\pi} x_i(0) - \frac{1}{2\pi} \dot{x}_i(0)$$

System of equations: $\begin{pmatrix} \omega^2 - 2\omega_0^2 & \omega_0^2 \\ \omega_0^2 & \omega^2 - 2\omega_0^2 \end{pmatrix} \cdot \begin{pmatrix} \tilde{x}_1 \\ \tilde{x}_2 \end{pmatrix} = \frac{1}{2\pi} \begin{pmatrix} i\omega x_1(0) - \dot{x}_1(0) \\ i\omega x_2(0) - \dot{x}_2(0) \end{pmatrix}.$

Second member with imposed initial conditions: $\frac{1}{2\pi} \begin{pmatrix} i\omega x_1(0) - \dot{x}_1(0) \\ i\omega x_2(0) - \dot{x}_2(0) \end{pmatrix} = -\frac{x_{10}}{2i\pi} \begin{pmatrix} \omega \\ 0 \end{pmatrix}.$

Resolution:

$$M^{-1} = \frac{1}{(\omega^2 - \omega_0^2)(\omega^2 - 3\omega_0^2)} \begin{pmatrix} \omega^2 - 2\omega_0^2 & -\omega_0^2 \\ -\omega_0^2 & \omega^2 - 2\omega_0^2 \end{pmatrix}$$

$$\rightarrow \begin{pmatrix} \tilde{x}_1 \\ \tilde{x}_2 \end{pmatrix} = -\frac{x_{10}}{2i\pi} \frac{\omega}{(\omega^2 - \omega_0^2)(\omega^2 - 3\omega_0^2)} \begin{pmatrix} \omega^2 - 2\omega_0^2 \\ -\omega_0^2 \end{pmatrix}$$

Inverse Laplace transform (integral with four poles):

$$x_1(t) = TL^{-1}[\tilde{x}_1(\omega)] = -\frac{x_{10}}{2i\pi} \int_C \frac{\omega(\omega^2 - 2\omega_0^2)}{(\omega^2 - \omega_0^2)(\omega^2 - 3\omega_0^2)} e^{-i\omega t} \, d\omega$$

$$= x_{10} \left[\frac{1}{4} e^{-i\omega_0 t} + \frac{1}{4} e^{i\omega_0 t} + \frac{1}{4} e^{-i\sqrt{3}\omega_0 t} + \frac{1}{4} e^{i\sqrt{3}\omega_0 t} \right]$$

$$= \frac{x_{10}}{2} (\cos \omega_0 t + \cos \sqrt{3}\omega_0 t)$$

$$x_2(t) = TL^{-1}[\tilde{x}_2(\omega)] = \frac{x_{10}}{2i\pi} \int_C \frac{\omega \omega_0^2}{(\omega^2 - \omega_0^2)(\omega^2 - 3\omega_0^2)} e^{-i\omega t} \, d\omega$$

$$= -x_{10} \left[-\frac{1}{4} e^{-i\omega_0 t} - \frac{1}{4} e^{i\omega_0 t} + \frac{1}{4} e^{-i\sqrt{3}\omega_0 t} + \frac{1}{4} e^{i\sqrt{3}\omega_0 t} \right]$$

$$= \frac{x_{10}}{2} (\cos \omega_0 t - \cos \sqrt{3}\omega_0 t)$$

Box 6.4. *Resolution by Laplace transformation*

The application of the Laplace transformation seems, in this simple case, a little more complicated than the previous methods, but it is nevertheless the easiest method to generalize (see section 6.8.2).

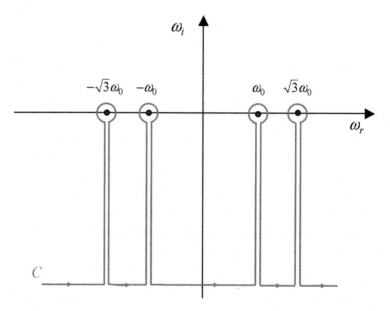

Figure 6.13. *Integration contour for transformation of inverse Laplace. For a color version of this figure, see www.iste.co.uk/belmont/plasma.zip*

6.8.2. *Laplace transformation for the calculation of kinetic modes*

Starting from the Vlasov–Maxwell system of equations and applying the Laplace transformation, we get this time the following system:

$$
\begin{cases}
(v - \omega/k)\tilde{f}_1 = \dfrac{ef_0'(v)}{ikm}\tilde{E}_1 + \dfrac{1}{2\pi}f_1^0(v) \\[2mm]
\tilde{E}_1 = -\dfrac{e}{ik\varepsilon_o}\tilde{n}_1, \text{ with } \tilde{n}_1 = \displaystyle\int dv\, \tilde{f}_1(v)
\end{cases}
$$

The spatial variation is given as e^{ikx}, with k being fixed. We have denoted by $E_1(t)$ the function of time and by $\tilde{E}_1(\omega)$ its Laplace transform. The same notations are used for f_1.

The function $f_1^0(v)$ is the initial perturbation value of $f_1(v)$. The function $f_0'(v)$ is the derivative with respect to the velocity of the zeroth-order distribution $f_0(v)$. We thus have:

$$\begin{cases} \tilde{f}_1 = \dfrac{e f_0'(v)}{ikm(v - \omega/k)}\tilde{E}_1 + \dfrac{1}{2\pi}\dfrac{f_1^0(v)}{v - \omega/k} \\ \tilde{E}_1 = \dfrac{e^2 \tilde{E}_1}{k^2 m \varepsilon_o}\displaystyle\int \dfrac{f_0'(v)}{v - \omega/k}dv - \dfrac{e}{2i\pi k \varepsilon_o}\displaystyle\int \dfrac{f_1^0(v)}{v - \omega/k}dv \end{cases}$$

Here, we divide by $v - \omega/k$ without any other precaution because we assume that ω is complex. This quantity is therefore never zero as long as the imaginary part ω_i is not zero.

We thus find:

$$\left[1 - \dfrac{e^2}{k^2 m \varepsilon_o}\int \dfrac{f_0'(v)}{v - \omega/k}dv\right]\tilde{E}_1 = -\dfrac{e}{2i\pi k \varepsilon_o}\int \dfrac{f_1^0(v)}{v - \omega/k}dv$$

We can write this as:

$$\tilde{E}_1 = -\dfrac{1}{D(\omega,k)}\dfrac{e}{2i\pi k \varepsilon_o}\int \dfrac{f_1^0(v)}{v - \omega/k}dv$$

where $D(\omega,k)$ is given by:

$$D(\omega,k) = 1 - \dfrac{e^2}{k^2 m \varepsilon_o}\int \dfrac{f_0'(v)}{v - \omega/k}dv = 1 - \dfrac{\omega_{pe}^2}{k^2}I(\omega/k)$$

with $I(\omega/k) = \displaystyle\int \dfrac{f_0'(v)/n_0}{v - \omega/k}dv$.

Any solution in E_1 is then calculated, for $t > 0$, as an inverse Laplace transform of this expression of $\tilde{E}_1$, which has the form of the following integral:

$$E_1(t) = \int_C \tilde{E}_1(\omega)e^{-i\omega t}d\omega = -\dfrac{e}{2i\pi k \varepsilon_0}\int_C \dfrac{1}{D(\omega,k)}\left(\int \dfrac{f_1^0(v)}{v - \omega/k}dv\right)e^{-i\omega t}d\omega$$

The general result thus has this form of an integral in ω, where the function under the integral sign must be estimated along a certain contour C, passing over all its poles. What are the poles of this function? We see that these are at least the zeros of D. In general, an infinite number of other poles can also come from the form of the initial perturbation $f_1^0(v)$ itself. Here, we again find the infinite number of possible "eigenmodes" solutions. We will assume here, as explained in this chapter, that $f_1^0(v)$ does not have poles in the complex plane, which implies that the initial condition is sufficiently "smooth" and "ordinary" (no special preparation). Under these conditions, this infinite number of solutions is restricted to the only solutions ω_n of $D(\omega, k) = 0$.

As explained in section 6.6.1, the effective form of $E_1(t)$ is then obtained by deformation of the integration contour and the result then appears as a simple sum of eigenmodes:

$$E_1(t) = \sum_n E_{1n} e^{-i\omega_n t}$$

where the values of ω_n are the solutions of a dispersion equation. $D_p(\omega_n, k) = 0$.

However, beware, this D_P is not exactly the denominator D that appears in the inverse Laplace transformation. There is a change to take into account when deforming the contour. On the original contour, we had:

$$D(\omega, k) = 1 - \frac{\omega_{pe}^2}{k^2} I(\omega/k) \text{ with } I(\omega/k) = \int \frac{f_0'(v)/n_0}{v - \omega/k} dv$$

This original contour (in the complex plane ω of the inverse Laplace transform) was necessarily above all the poles ω_n of interest. We can therefore assume $\omega_i > 0$ throughout this contour. The integral giving I, which is an integral on real velocities, posed no problem of definition (no real pole) for ω belonging to this contour.

On the contrary, when we deform the contour to isolate the role of the different poles, there is a problem that every time the modes are damped, that is, for $\omega_{ni} < 0$ (this is the case for all solutions for a stable medium). Indeed, when we pass to $\omega_i = 0$ during the deformation, the integral I has a pole that makes it indefinite and, then, its value for $\omega_i < 0$ does not continually continue the values at $\omega_i > 0$. According to the integration theory in the complex plane, the deformation is only valid for a holomorphic function, which means that for $\omega_i < 0$, the integral I must be replaced by its analytical continuation I_P. This can be done for example by replacing the real variable v by a complex variable and by deforming the integration contour of I for this v complex. We thus find, for the extended function I:

$$D_P(\omega, k) = 1 - \frac{\omega_{pe}^2}{k^2} I_P(\omega / k)$$

with: $I_P(\omega / k) = VP \int dv \dfrac{f_0'(v) / n_o}{v - \omega / k} + 2i\pi \; \alpha \; f_0'(\omega / k) / n_o.$

and:

- $-\alpha = 0$ for $\omega_i > 0$;
- $-\alpha = 1/2$ for $\omega_i = 0$;
- $-\alpha = 1$ for $\omega_i < 0$.

7

Shockwaves and Discontinuities

The shockwave is a well-known phenomenon in neutral gases. We know that it is produced, for example, in front of an aircraft flying at a speed higher than the speed of sound relative to the air. This phenomenon can also occur in plasmas in a very similar way. Thanks to the coupling that exists between matter and electromagnetic fields, a great variety of discontinuities can exist: several kinds of shocks, as well as other types of discontinuities that are not shocks, like rotational discontinuities. All of these discontinuities have in common the constitution of thin boundaries, which separate two regions of different characteristics, all parameters experiencing a jump between the upstream and downstream sides of the discontinuity.

7.1. Some examples of shocks and discontinuities

Figure 7.1 shows the interaction between a solid obstacle and a moving fluid. When the relative speed between the fluid and the obstacle is not very high, the fluid bypasses the obstacle (top). In the opposite case (bottom), a shockwave is formed at the front. What differentiates the two situations is the ratio of this relative speed to the speed of sound. Why the speed of sound? Because it is the speed of propagation of pressure fluctuations in the fluid. We can summarize the situation by saying that in the subsonic case, the information of the existence of the obstacle can propagate to any element of the fluid upstream, as far as it can be, even if the corresponding pressure fluctuation gradually decreases. In the supersonic case, the situation is very different: beyond a certain distance upstream of the obstacle, the fluid completely "ignores" the existence of the obstacle. The fluid flow is only disturbed by the presence of the obstacle very close to it, downstream of a thin boundary: the shock.

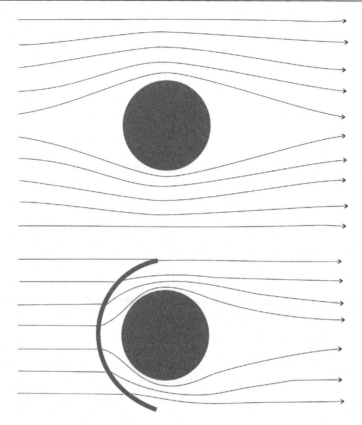

Figure 7.1. *Diagram illustrating the interaction of a fluid with an obstacle, depending on whether the fluid velocity is subsonic (top) or supersonic (bottom)*

Shocks are observed in neutral fluids, such as in front of a missile (see Figure 7.2). In this figure, it can be seen that the characteristic properties that define a shock can be encountered in very different environments and that the speeds involved can be very different. In the case of a duck swimming on water, obviously the speed of sound does not characterize the shock: the duck is rarely supersonic (remember that the speed of sound in the atmosphere is of the order of 1,000 km/h)! The information to be propagated in this case is not a change in air pressure but a variation in the height of the water's surface. The speed of propagation for this information is the speed of the "waves", which is much lower. It is a dispersive wave whose propagation speed depends on the wavelength, but for the wavelengths usually excited by a duck, the order of magnitude of this speed is about 1 or 2 km/h. It is thus enough for the duck to exceed this velocity to produce the characteristic structure of the shock and the wake that follows it.

Figure 7.2. *In both the example of the duck and that of the missile (test carried out in a wind tunnel, NASA Langley Research Center), we see common characteristics: undisturbed environment far upstream, borders close to the obstacle, etc. The reason for the missile is that it is supersonic. How about for the duck?*

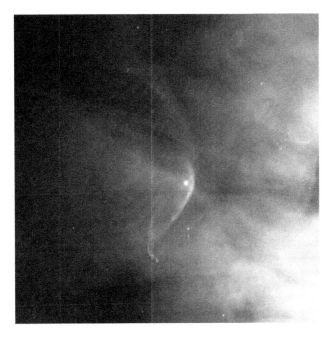

Figure 7.3. *The Orion Nebula (source: NASA and the Hubble Heritage Team [STScI/AURA]). For a color version of this figure, see www.iste.co.uk/belmont/plasma.zip*

Shocks are also observed in the plasmas, and Figure 7.3 shows an example of astrophysical shock observed in the Orion Nebula. The bright brilliant arc is a cosmic shock that is about half a light-year wide. In three dimensions, the structure has the shape of a bowl. In this context too, this type of structure occurs when we have two backgrounds with very different speeds that meet. Here, the plasma is not completely ionized: the collisions with the neutral atoms give the shock its luminous character. It is said that it is a "radiative" shock.

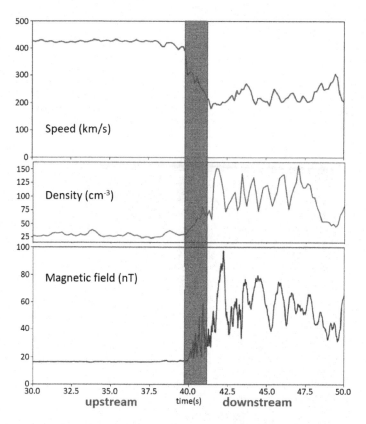

Figure 7.4. *Observations of the MMS1 satellite (NASA) on October 7, 2015, starting from 11:44:30. The three quantities represented show a sudden change in properties. At the beginning of the time interval, the speed is very high; it is the speed of the solar wind. As the shock passes (underlined in blue), it is divided by 2. At the same time, we observe a compression of the plasma and an increase in the magnetic field. For a color version of this figure, see www.iste.co.uk/belmont/plasma.zip*

Much closer to us, there is a similar structure. The solar wind emitted by the Sun is brutally slowed down by a shockwave before reaching the terrestrial magnetosphere. This shockwave is not observable in visible light like the previous one (the plasma is completely ionized and "collisionless"), but it has been evidenced by the measurements made by the different satellites that are in orbit around the Earth. Figure 7.4 shows an example of measurements that clearly show the existence of a shock: a jump in the value of the density and in the value of the magnetic field is observed.

It is known that, in a neutral fluid, behind the shock, there is an area of strong turbulence. In plasma, we observe the same thing. In the shock and downstream (Figure 7.4), there is a high level of magnetic fluctuation.

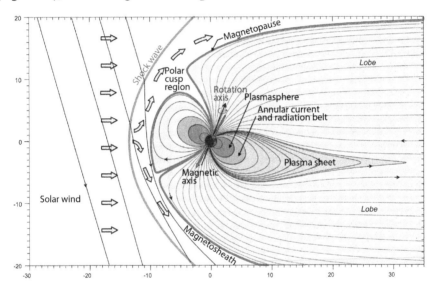

Figure 7.5. *Diagram showing the Earth's magnetosphere. The fine lines represent the magnetic field lines. We have to imagine the Sun very far to the left; the solar wind carries the magnetic field lines related to the Sun. Starting from the left, we first encounter the shockwave and then the magnetopause, before the internal regions of the magnetosphere. For a color version of this figure, see www.iste.co.uk/belmont/ plasma.zip*

Figure 7.5 shows a schematic view of the Earth's magnetosphere. We see that, coming from the Sun (on the left), we meet the shock and then a second boundary, the magnetopause. As its name suggests, it is a magnetic boundary: it is the separation between the magnetic field lines that are related to the Earth and the field lines that are related to the interplanetary medium. The direction of the latter is mainly determined by that of the magnetic field in the solar wind, even if they are slightly compressed by the shock. The magnetic field is mainly determined, on the contrary, by the terrestrial dipole. This orientation slightly depends on the day and the hour because the terrestrial magnetic dipole is not aligned with the Earth's rotation axis and is not perpendicular to the ecliptic plane, but it is not directly related to the solar magnetic field. We therefore generally observe a rotation of the field at the magnetopause. There is also a strong variation in density, the magnetosphere being substantially less dense than the incident plasma (the magnetosphere is a "bubble" in the flow of the solar wind). This second boundary, coming from the Sun, is also a thin current layer, like the shock, but it will be seen later that it is a discontinuity of a nature different from that of a shock.

Figure 7.6 shows that shocks can also be produced in laboratory plasmas. They are most often created by laser or a plasma jet, and sometimes they are created to reproduce the shocks observed in astrophysics.

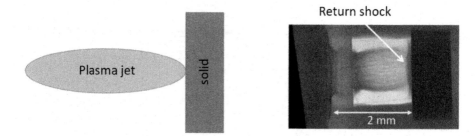

Figure 7.6. *Laboratory observation of shock formation by interaction of a plasma jet with a solid: (left) diagram of the experiment; (right) image obtained by X-ray radiography (source: A. Benuzzi-Mounaix)*

7.2. Existence of discontinuities

7.2.1. *Absence of stationary gradient*

We first consider a neutral adiabatic fluid (we suppose p/ρ^γ = cste, with γ = 5/3). It is supposed that this fluid is moving towards the positive values of x. If an obstacle is placed in this stream, then what stationary form can the flow around this obstacle take? As the fluid's velocity v_x has to cancel out on the obstacle, we look, in particular, under what conditions this slowing down, between the limit condition on the left $v_x = v_o$ and the limit condition on the right $v_x = 0$, can be continuous and when it must be done through a discontinuity.

By simplifying this problem to one dimension (x), several interesting results can be established in a simple way. We start with the system of fluid equations:

$$\begin{cases} \partial_t \rho + \partial_x(\rho v_x) = 0 \\ \rho\partial_t v_x + \rho v_x\partial_x(v_x) + \partial_x(p) = 0 \end{cases}$$

Taking into account the equation linking p and ρ and the definition of the speed of sound $c_s = \sqrt{\dfrac{5}{3}\dfrac{p}{\rho}}$, we can prove the following differential equations: $\partial\rho = 3\rho\partial c_s/c_s$ and $\partial p = 5p\partial c_s/c_s$.

The system then goes into the following form:

$$\begin{cases} 3\left(\partial_t + v_x\partial_x\right)\left(c_s\right) + c_s\partial_x\left(v_x\right) = 0 \\ \left(\partial_t + v_x\partial_x\right)\left(v_x\right) + 3c_s\partial_x\left(c_s\right) = 0 \end{cases} \qquad [7.1]$$

Let us first find out if there are "stationary" solutions to the above system (partial derivative with respect to time equal to zero in some reference frame). It will be admitted that these solutions can be stationary in any reference frame: if this frame moves at speed V with respect to the obstacle, then the relative speed of the plasma relative to the structure can be determined as: $\delta v_x = v_x - V$. With this notation, we have $\partial_t = - V\partial_x$, and the system is written as:

$$\begin{cases} 3\delta v_x\partial_x\left(c_s\right) + c_s\partial_x\left(\delta v_x\right) = 0 \\ \delta v_x\partial_x\left(\delta v_x\right) + 3c_s\partial_x\left(c_s\right) = 0 \end{cases}$$

The resolution in δv_x then simply leads to $\left[\delta v_x^2 - c_s^2\right]\partial_x\left(\delta v_x\right) = 0$. Now, it is easy to see that this equation has no exact continuous solution other than the trivial solution δv_x = cste and c_s = cste (uniform). This nonlinearly exact result, in which the velocity is uniform, is obviously incompatible with the hypothesis that the boundary conditions are different on the right and on the left. The conclusion is therefore that it is not possible to slow the flow upstream of the obstacle by a continuous gradient that is stationary (in its reference frame).

7.2.2. *Nonlinear steepening*

If, therefore, a one-dimensional gradient can never be stationary, then it means that any gradient taken as an initial condition will, in addition to its global propagation, either spread (and tend towards the uniform solution) or steepen (and tend towards a discontinuity).

We can understand this phenomenon by studying a particular case called "simple wave", which is a nonlinear extension of the notion of progressive wave. It suffices to consider variations of v_x and c_s, which are connected to each other by $v_x + 3c_s =$ constant. By summing equations [7.1], we verify that:

$$d_{t+}(v_x + 3c_s) = 0, \text{ where } d_{t+} = \partial_t + (v_x + c_s)\partial_x.$$

Therefore, if this condition is verified everywhere in the initial condition, it continues to be so over time. Under these conditions, the difference of the initial equations shows that the propagation then respects:

$$d_{t-}(v_x) = 0 \text{ and } d_{t-}(c_s) = 0, \text{ with } d_{t-} = \partial_t + (v_x - c_s)\partial_x.$$

This means that if we give a value $v_x = v_{xo}$ at a time $t = t_o$, then we find this value at time $t = t_o + \delta t$ at the point $x = x_o + (v_x - c_s)\,\delta t$. Thus, we can simply deduce the evolution of a given profile chosen as an initial condition. The same reasoning applies of course for the profile of c_s.[1]

Figure 7.7 illustrates the evolution of an initially sinusoidal profile that satisfies $v_x + 3c_s =$ cste. We see that the gradients in the direction of propagation steepen, while the opposite sides flatten. It can also be noted that the relative speed of propagation with respect to the plasma is equal to c_s everywhere (in absolute value).

It should be noted that the linear result indicates the average propagation speed of the structure $< v_x - c_s > = v_{x0} - c_{s0}$, whereas the complete nonlinear computation makes it possible to account for the variations with respect to this average speed, leading to the deformation of its profile and, in particular, to the steepening of certain gradients.

---

1 The calculation presented here, which uses the notion of "simple wave", is an elementary example of a more general mathematical method, called the characteristic method. This is used to solve the problems of nonlinear propagation (see Witham, 1974), that is, to solve the hyperbolic systems of partial differential equations.

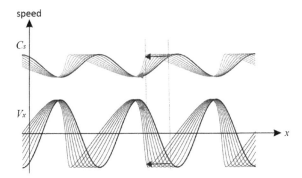

Figure 7.7. *The curves represent the speed profile for a given time (sinusoidal profile at the initial time, the rightmost, with increasingly clearer colors when t increases). The arrows indicate the displacement* $(v_x - c_s)\delta t$ *proportional to the propagation speed:* $v_x - c_s$. *This is larger in absolute value where* v_x *is more negative. For a color version of this figure, see www.iste.co.uk/belmont/plasma.zip*

COMMENT ON FIGURE 7.7.– If, at time t_0, we know the speed of the fluid v_{x0} and the speed of sound c_{s0} at position x_0, then we know that at time $t_0 + \delta t$, we find these same values v_{x0} and c_{s0} at position $x_0 + (v_x - c_s)\,\delta t$. This causes the steepening of the front edge of the wave during its propagation (to the left in the reference frame of the figure).

7.2.3. *Formation of discontinuities*

If we pushed the above construction further, we would arrive at an absurd result: the profile would come to "break", that is, in certain regions we would find three different values for the speed v_x at the same point x^2. We are confronted with this problem whenever, at certain points, we can deduce the value of the speed from several different initial values and the different values found are contradictory. It can be said, very generally, that this type of contradiction between boundary conditions (spatial or temporal) is always the cause of the existence of discontinuities. When an obstacle is placed in a supersonic flow, for example, the ideal propagation, in stationary regime, would impose that the velocity given at the upstream limit remains constant everywhere downstream (no information goes fast enough to "go up" the flow); this is obviously in contradiction with the presence of the obstacle that imposes a zero velocity on its surface, and hence a bow shock must necessarily exist before the obstacle.

---

2 This would not be impossible if we studied the case of a real wave breaking on water, that is, if the calculated quantity was the level of the water during the propagation of such a wave: this level can actually have three different values at the same point. On the contrary, a multi-value solution is obviously unacceptable for a quantity such as the fluid's speed.

What limits the steepening of a profile and can oppose the aforementioned breaking? What happens in practice is that the "ideal" fluid equations that have just been solved cease to be valid when the steepening leads to gradients that are too large (e.g. because of the viscosity, even though it can be very small). These non-ideal effects "stationarize" the profile: the shock in these conditions appears as a balance between nonlinear (steepening) effects and non-ideal effects (which oppose the steepening). It is only because of the non-ideal effects that the propagation can reach values different from c_s (local speed of sound) inside the structure and be brought to a supersonic speed upstream of a shock and subsonic downstream.

Let us note that the discontinuities thus formed are boundaries which are all the finer as non-ideal effects, such as viscosity, are weak. This will justify the one-dimensional hypothesis used below to study them locally. On the contrary, three-dimensional effects are generally important for studying the behavior of the rest of the fluid. We have considered here a one-dimensional flow everywhere only to simplify the presentation: this can lead to significant differences in the overall result. For example, the flow around a spherical obstacle leads to a shockwave immobile in the reference frame of the obstacle, whereas a plane obstacle leads to a shockwave that moves away from the obstacle (there is an accumulation of material behind the shock since the flow has no escape).

Let us finally note that, if the demonstrations were presented here for the simple case of a neutral fluid, the phenomena are quite comparable for any other fluid and, in particular, in MHD. The rest of the chapter is devoted to MHD discontinuities.

7.3. Establishment of jump equations

7.3.1. *Definition of jump equations*

A discontinuity, by definition, is a stationary and one-dimensional structure (the quantities vary essentially only according to the normal **n**). It should be noted that this definition does not assume that the boundary is thin, except, possibly, with respect to the scale characterizing the deviation from plane geometry (e.g. curvature radius). For calculations, we will place ourselves in the reference frame where the structure is stationary.

It will be further assumed that the different parameters reach constant values asymptotically on either side of the discontinuity (Figure 7.8). In this way, we can unambiguously define, for any quantity A, its asymptotic values upstream and downstream of the structure A_1 and A_2, as well as the jump of this quantity $\Delta A = A_2 - A_1$. The jumps of the different parameters are connected to each other by the "jump equations". These are also called the "Rankine–Hugoniot equations" in a neutral gas and "generalized Rankine–Hugoniot equations" in MHD. We are not interested here in what happens inside the discontinuity.

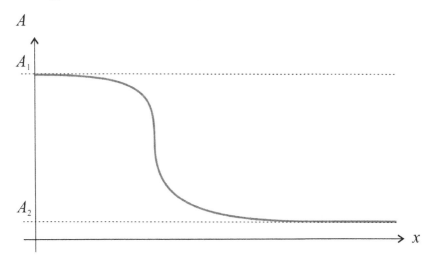

Figure 7.8. *Evolution of a quantity at the passage of the discontinuity*

The different jump equations are derived from the conservation equations of the system of equations considered, which is, in this case, the MHD system. The equations of the ideal MHD come from Maxwell's equations, on the one hand, and from fluid conservation equations, on the other hand: mass, momentum and energy (see Chapter 5). It must be remembered that the latter have the simple form known to them, only subject to a simplifying hypothesis: the isotropy of the distribution function (*p* scalar pressure). This hypothesis can easily be accepted outside the boundary layer, but *a priori* not inside it: the viscosity, for example, is based on the existence of non-diagonal terms in the pressure tensor. Nevertheless, we will retain the isotropic shape of the conservation equations, even inside the layer, because the differences introduced by this approximation of the results remain, in general, negligible when the layer is sufficiently thin. On the contrary, we will not make any hypothesis of the "closure equation" type to express the heat flux **q** inside the layer, contrary to what is conventionally used to solve the MHD system (adiabatic, isothermal, polytropic, etc.). We will see that, in order to solve it, it will suffice to assume the values of **q** outside the layer, the simplest naturally being **q** = 0 on both sides. The character, being always non-adiabatic, of phenomena inside the layer indeed plays an essential role for the existence of certain discontinuities. The results that we are going to demonstrate in this way therefore have broader validity than MHD.

All the MHD equations, including the energy equation (as long as it is complete, i.e. including the heat flux), can be put in a conservative form (see section 4.6.1)[3]:

$$\partial_t(a) + \nabla.(\mathbf{b}) = 0$$

Since we are looking for stationary and one-dimensional solutions of equations, the equations take the form:

$$\nabla.(\mathbf{b}) = 0 \Rightarrow \mathbf{n}.\partial_x(\mathbf{b}) = 0 \Rightarrow \partial_x(b_n) = 0$$

where $\mathbf{n}$ is the normal to the discontinuity and b_n is the component of $\mathbf{b}$ along this normal. These equations are easily integrated and give:

$$\mathbf{n}.(\mathbf{b}_2 - \mathbf{b}_1) = 0 \Leftrightarrow \mathbf{n}.\Delta\mathbf{b} = 0 \Leftrightarrow \Delta b_n = 0$$

(Warning: in the whole chapter, Δ indicates a jump; it has no relation to a Laplacian.) Each equation of the MHD system thus makes it possible to establish a conservation equation at the crossing of the discontinuity.

7.3.2. Application to the MHD equations

The integration made on a quantity a (see section 7.3.1) can now be applied to all MHD equations. The conservation equation of the mass makes it possible to establish that:

$$\rho_2 u_{2n} = \rho_1 u_{1n} = \Phi_m$$

where the index n indicates the normal component of the velocity. In the same way, $\nabla.(B) = 0$ implies the well-known conservation of the normal component of the magnetic field at the crossing of the discontinuity:

$$B_{2n} = B_{1n} = B_n$$

The conservation equations of momentum and energy are a little heavier to handle; they allow us to establish the conservation of the quantities:

$$\rho_2 u_{n2}\mathbf{u}_2 + \left(p_2 + \frac{B_2^2}{2\mu_0}\right)\mathbf{n} - \frac{B_n\mathbf{B}_2}{\mu_0} = ... = \Phi_i$$

_____

3 When a is a scalar, $\mathbf{b}$ is a vector; when $\mathbf{a}$ is a vector, $\mathbf{b}$ is a tensor of order 2 and so on.

$$\frac{1}{2}\rho_2 u_2^2 u_{n2} + \frac{5}{2}p_2 u_{n2} - \frac{1}{\mu_0}\left[B_n\left(\mathbf{B}_2.\mathbf{u}_2\right) - B_2^2 u_{n2}\right] = \ldots = \Phi_e$$

(The right-hand side is identical to the left-hand side, with indices 1.) For the energy equations, it should be noted that heat fluxes do not appear: indeed, although they have been assumed to be non-zero inside the layer, the fact that they are null outside of it is enough to make them disappear from the jump equations. Finally, the Maxwell–Faraday equation $\partial_t(\mathbf{B}) = -\nabla\times(\mathbf{E})$ leads to the conservation of the tangential component of the electric field $\mathbf{E}_{T1} = \mathbf{E}_{T2} = \mathbf{E}_T$, which is written, considering that the ideal Ohm's law is supposed to be verified on both sides:

$$u_{n2}\mathbf{B}_{T2} - B_{n2}\mathbf{u}_{T2} = u_{n1}\mathbf{B}_{T1} - B_{n1}\mathbf{u}_{T1} = \mathbf{n}\times\mathbf{E}_T$$

7.3.3. de Hoffmann–Teller frame

It has already been said that the calculation reference frame is chosen so that the discontinuity is immobile. This fixes the motion of the reference frame according to the normal; however, since the problem posed is invariant for any translation in the tangent plane, there remains a degree of freedom as regards the tangential motion of the reference frame used. We will generally choose the de Hoffmann–Teller frame, named after its authors (de Hoffmann and Teller 1950). By definition, this reference is that in which the velocity is parallel to the magnetic field; that is, thanks to the ideal Ohm's law, the electric field is zero. The fact that $\mathbf{E} = 0$ everywhere greatly simplifies the study of discontinuities.

To demonstrate that such a reference exists, it must first be noted that in each medium (upstream or downstream), the fact that the tangential component of the electric field is zero is generally sufficient to cause the complete electric field vector to be zero:

$$E_T = 0 \Rightarrow E = 0$$

Indeed, the use of the ideal Ohm's law shows that $\mathbf{E}_T = \mathbf{n}\times(B_n\mathbf{u}_T - u_n\mathbf{B}_T)$. Therefore, when $\mathbf{E}_T = 0$, $\mathbf{u}_T$ and $\mathbf{B}_T$ are necessarily collinear (so long as u_n and B_n are non-zero), which means that $\mathbf{E}_n = -\mathbf{u}_T\times\mathbf{B}_T$ is zero too.

We know that the electric field can be cancelled on one side of the discontinuity (upstream or downstream) by a simple change of reference frame at velocity

$\dfrac{\mathbf{E}}{B}\times\mathbf{b}$ (see Chapter 2). This amounts to adding to the original plasma velocity the velocity necessary to align the resulting velocity with the magnetic field. Since the tangential component of the electric field $\mathbf{E}_T$ is preserved at the crossing of the discontinuity, the cancellation of the tangential electric field on one side necessarily leads to its nullity on the other side. According to the previous remark, the complete electric fields are therefore null on both sides in the reference frame so determined.

There is, however, a case where we cannot find a de Hoffmann–Teller frame: it is that of strictly perpendicular discontinuities $B_n = 0$ where the field is tangential to the discontinuity. In these circumstances, two cases may arise:

– $u_n \neq 0$: we can cancel E_n by a tangential change of the reference frame, but this leaves $\mathbf{E}_T$ unchanged and therefore not zero. $\mathbf{E}_T$ = cst imposes that $\mathbf{B}_T$ remains parallel to itself and that its modulus varies as $1/u_n$ (it is simply the limit of the oblique fast shock that we will find in the following section when the direction $\mathbf{n}$ tends towards the direction perpendicular to $\mathbf{B}$);

– $u_n = 0 \Rightarrow \mathbf{E}_T = \mathbf{0}$, but E_n is, in general, neither zero nor preserved at the crossing (tangential discontinuity).

7.4. Different types of discontinuities that can exist in plasma

7.4.1. System solutions

We place ourselves in the de Hoffmann–Teller frame defined above, and we omit the indices 2 for ease of writing. The jump equation deduced from the conservation of momentum equation can be projected on the normal and tangential directions, which gives:

$$\rho u_n^2 + p + \dfrac{B_T^2}{2\mu_0} = \Phi_{in}$$

$$\rho u_n \mathbf{u}_T - \dfrac{B_n \mathbf{B}_T}{\mu_0} = \mathbf{\Phi}_{iT}$$

The second equation can be simplified by using the de Hoffmann–Teller equation $\mathbf{u}_T = \dfrac{u_n}{B_n}\mathbf{B}_T$ (coming from $\mathbf{E}_T = 0$) and put in the form:

$$\left(u_n - \dfrac{B_n^2}{\mu_0\Phi_m}\right)\mathbf{B}_T = \dfrac{B_n}{\Phi_m}\mathbf{\Phi}_{iT} \Rightarrow \left(u_n - \dfrac{B_n^2}{\mu_0\Phi_m}\right)\mathbf{B}_T = \left(u_{n1} - \dfrac{B_n^2}{\mu_0\Phi_m}\right)\mathbf{B}_{T1}$$

This equation shows that there are two types of rather different solutions:

– the tangential magnetic field keeps its direction at the passage of discontinuity and it is determined by the equation above. The discontinuity thus obtained is said to be coplanar, since the fields, and thus also the velocities in the de Hoffmann–Teller frame, are in the same plane upstream and downstream[4]. We can refer to "refraction of the field lines" (see Figure 7.9). The common plane is determined by the normal and the direction of the incident magnetic field ($\mathbf{B}_{T1}$, $\mathbf{B}_T$, $\mathbf{u}_{T1}$, $\mathbf{u}_T$ being all collinear). This type of solution is called *shock*;

– A second solution of the above equation consists of canceling the two parentheses, which eliminates the constraint on the tangential field. In this case, the tangential field can rotate and the discontinuity is called rotational (Figure 7.10).

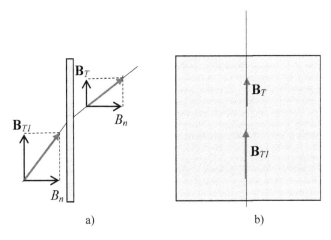

a)   b)

Figure 7.9. *Evolution of the magnetic field at the crossing of a shockwave (coplanar discontinuity), a) in projection in the plane perpendicular to the discontinuity and b) in the plane of the discontinuity. The upstream field is shown in red, and the downstream field in green. For a color version of this figure, see www.iste.co.uk/belmont/plasma.zip*

4 NOTE.– vectors on both sides $\mathbf{B}_1$ and $\mathbf{B}$ always belong to a plane (and all intermediate values if the polarization of $\mathbf{dB} = \mathbf{B} - \mathbf{B}_1$ is linear). What distinguishes the "coplanar" case from the general case is that this plane contains the normal. As seen in Figure 7.9, the shape of a field line then resembles that of a refracted light ray on a flat surface: the outgoing ray is in the plane defined by the incoming ray and the normal.

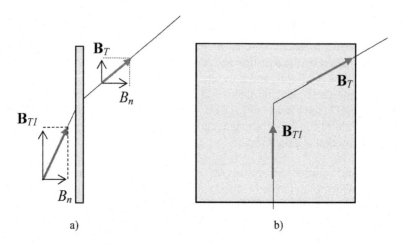

a)   b)

Figure 7.10. *Evolution of the magnetic field at the crossing of a rotational discontinuity, (a) in projection in the plane perpendicular to the discontinuity and (b) in the plane of the discontinuity. The upstream field is shown in red, and the downstream field in green. For a color version of this figure, see www.iste.co.uk/belmont/plasma.zip*

7.4.2. *Properties of rotational discontinuities*

As we have just seen, we have in this case:

$$u_n - \frac{B_n^2}{\mu_0 \Phi_m} = u_{n1} - \frac{B_n^2}{\mu_0 \Phi_m} = 0 \Rightarrow u_n = u_{n1} = \frac{B_n^2}{\mu_0 \Phi_m}$$

Since all quantities that appear in this expression, apart from u_n, are conserved, the normal component of the velocity has to be conserved, too. The conservation equation of the normal momentum, therefore, implies that the density is conserved as well. It is therefore a discontinuity that is not accompanied by a slowing down of the plasma or compression, which is very different from shock.

Under these conditions, the conservation equation of the momentum projected in the direction of the normal causes the conservation of $p + \frac{B_T^2}{2\mu_0}$. In fact, p and B_T^2

are actually conserved separately because the conservation equation of energy (by simplifying the terms that are obviously conserved) implies the conservation of:

$$\rho \frac{v_T^2}{2} u_n + \frac{5}{2} p u_n - \frac{1}{\mu_0} \left(B_n \left(\mathbf{B}_T . \mathbf{u}_T \right) - B_T^2 u_n \right)$$

By replacing the tangential velocity with its expression as a function of the tangential field, we find an expression that depends only on the squared modulus of the tangential field B_T^2; the latter is thus conserved, too. This furthermore justifies the name of the rotational discontinuity, since the tangential component of the magnetic field merely rotates without changing the modulus. The same goes for the tangential speed (in the de Hoffmann–Teller frame).

We will remember that, in a rotational discontinuity, the following quantities are preserved:

$$\boxed{u_n,\ \rho,\ p, B_T,\ u_T}$$

It is also worth remembering that the rotational discontinuity propagates, with respect to the plasma, at the Alfvén speed along the magnetic field: it is nothing else but the nonlinear form of a shear Alfvén wave.

7.4.3. Shocks

Shocks are characterized by the coplanarity of the magnetic field (and velocity in the de Hoffmann–Teller frame). The resolution of the system, eliminating all variables except the normal velocity, leads to an equation in u_n, whose solution is shown in Figure 7.10. For very large values of the normal incident speed u_{nl} (more to the right than the part of the curve shown in the figure), the curve goes back to an oblique asymptote of slope less than 1.

A first solution is the trivial solution, $u_n = u_{nl}$: it corresponds to the absence of discontinuity, which is always a solution of the initial system. The other solutions that have a physical meaning are those which have the effect of reducing the normal speed (the lost energy being dissipated in the discontinuity) and which therefore correspond to $u_n < u_{nl}$. We thus note that, on the one hand, all the incident velocities do not give a solution and, on the other hand, there are three families of solutions, and we will see that they correspond to discontinuities with different properties.

The different types of solutions are separated by the particular speeds V_{s1}, V_{i1} and V_{f1}, indicated in the figure, which correspond to the intersections of the curve with the first bisector. The resolution of the equation makes it possible to identify these particular speeds as being the phase velocities in medium 1, respectively of the slow mode (V_{s1}), of the Alfvén mode (or "intermediate" mode: $V_{i1} = V_{An1}$) and of the fast mode (V_{f1}) for a wave that would propagate in the direction of the normal at the discontinuity. For the fast and slow modes, this result can be interpreted as follows: for a point very close to one of these points of intersection; u_n tends towards u_{n1}; the difference between the two becomes small; we can linearize the equations and it is therefore normal to find the solutions of the linear system.

The solutions are therefore denominated from the corresponding linear solutions:

– the *slow shock*, for $V_{s1} < u_{n1} < V_{i1}$;

– the *fast shock*, for $V_{f1} < u_{n1}$.

For $V_{i1} < u_{n1} < V_{s1max}$, we call the solution an intermediate shock (V_{s1max} being the upper value corresponding to the vertical tangent of the curve). Velocity V_{i1} is the phase velocity of the Alfvén mode, but it is important to note that the limit of the intermediate shock when u_{n1} tends to V_{i1} is not the linear Alfvén mode. The Alfvén mode is the linear limit of the rotational discontinuity and not of the intermediate shock (the polarization of this mode is rotational and not coplanar; see Chapter 5). The reason why the frequency V_{i1} of the Alfvén mode appears in the computation of the coplanar discontinuities is that the rotationality and the coplanarity are not strictly contradictory properties: it is enough that the rotation of $\mathbf{B}_T$ is done exactly on one half-turn (or one integer number of half-turns) so that the downstream field is in the same plane as the upstream field. It will be verified, in fact, below that, for an intermediate shock, $\mathbf{B}_T$ and $\mathbf{u}_T$ reverse (change of sign); this reversal is therefore clearly beyond the scope of linear variations. It is worth noting that the intermediate shock therefore always constitutes, even when u_n undergoes an infinitesimal jump, a nonlinear solution.

As we have just seen, each shock (slow, intermediate, fast) must have, in order to exist, an incident velocity u_{n1} greater than a characteristic velocity (V_{s1}, V_{i1}, V_{f1}). We could also check that the normal velocity downstream u_{n2} is each time smaller than the corresponding characteristic velocity (v_{s2}, v_{i2}, v_{f2}). Here, we find the equivalent of a well-known result for neutral gases: the normal velocity is supersonic upstream and subsonic downstream. For incident velocities belonging to none of the three ranges defined above, there is no possible discontinuity.

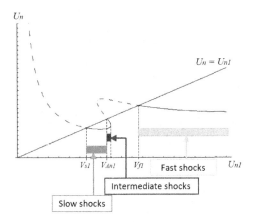

Figure 7.11. *Solutions of the equation in U_n for coplanar discontinuities (shocks). (Parameters chosen for the calculation: $\theta = 30°$, $\beta = 0.8$). For a color version of this figure, see www.iste.co.uk/belmont/plasma.zip*

The three types of shocks identified above have different properties that can be accessed by completing the system resolution. One of the distinguishing features is the variation of the tangential magnetic field. The fields on both sides of the shock are connected by:

$$\left(u_n - \frac{V_i^2}{u_n} \right) \mathbf{B}_T = \left(u_{n1} - \frac{V_{i1}^2}{u_{n1}} \right) \mathbf{B}_{T1}$$

noting: $V_{i1}^2 = \dfrac{B_n^2}{\mu_0 \rho_1} = \dfrac{B_n^2}{\mu_0 \Phi_m} u_{n1}$ and $V_i^2 = \dfrac{B_n^2}{\mu_0 \rho} = \dfrac{B_n^2}{\mu_0 \Phi_m} u_n.$

We can also note:

$$\mathbf{B}_T = F(u_n)\mathbf{B}_{T1}$$

with:

$$F(u_n) = \frac{u_{n1} - \dfrac{V_{i1}^2}{u_{n1}}}{u_n - \dfrac{V_i^2}{u_n}}$$

The factor F determines the variation of the magnetic field. Its study shows that, for intermediate shocks, it is negative, which means that there is a reversal of the tangential magnetic field at the passage of the discontinuity. The tangential velocity is connected to the tangential magnetic field by a factor that does not change sign, and therefore undergoes the same reversal. For fast and slow shocks, F is positive and the tangential fields on both sides are in the same direction. With respect to the $\mathbf{B}_T$ module, it increases or decreases depending on whether the factor F is, in absolute value, greater or smaller than 1. It can be shown that it is smaller than 1 for intermediate and slow shocks (decrease in magnetic energy) and greater than 1 for fast shocks (increase in magnetic energy).

We finally summarize the properties of shocks in Table 7.1.

Slow shock	Intermediate shock	Fast shock
Decrease in u_n, B_T Increase in ρ, p	Decrease in u_n, B_T Increase in ρ, p Reversal of B_T, u_T	Decrease in u_n Increase in ρ, p, B_T

Table 7.1. *Properties of the three types of shocks in magnetized plasma*

7.4.4. *Tangential discontinuity*

This is a very special case where the normal magnetic field is zero, at the same time as the normal velocity. This removes any relationship between the tangential magnetic fields on both sides, which are thus decoupled. The same goes for the tangential velocities. Only the conservation of the normal component of the momentum still provides a relationship between upstream and downstream. It results in the conservation of $p + B^2/2\mu_o$, that is, the total pressure (sum of the kinetic and magnetic pressures).

NOTES.–

– If $u_n = 0$ but $B_n \neq 0$, then almost all of the variables of the system are necessarily conserved. The only non-trivial solution is thus a solution where only the density and the temperature vary, but in such a way that their product, the pressure, remains constant. This is called a "contact discontinuity".

– The inverse hypotheses $B_n = 0$ but $u_n \neq 0$ characterize a fast shock in perpendicular propagation; no singular property is attached to it.

– It is found that much of the discussion about the nature of the discontinuity is based on whether the normal components of the magnetic field and velocity are null or not. This poses an experimental problem: it is not always easy to know whether a measurement is null or simply small. The questions of measurement uncertainty necessarily come into play.

7.5. Magnetospheric boundaries

We have seen in the first part of section 7.1 that the Earth's magnetosphere is separated from the solar wind by two boundaries. The first is clearly a fast shock. The second, the magnetopause, is a boundary that involves both a rotation of the magnetic field and a compression (modulus of the magnetic field and density of the plasma). It cannot be either a pure rotational discontinuity (there would be no compression) or a slow shock (there would be no rotation). We can conclude that in the stationary state the magnetopause can only be a tangential discontinuity, with $B_n = 0$ and $u_n = 0$. This point is difficult from the experimental point of view: if the difference between the oblique discontinuities and the tangential discontinuity is very clear on the theoretical plane, it is not really the case from this point of view. The normal component of the magnetic field is always very small compared to the tangential component. To be able to affirm that the component B_n is null or not, it is necessary to control the experimental uncertainties, in particular to determine the normal to the discontinuity in a precise way, by supposing that this normal is the same all along the crossing. This only makes sense in the theoretical case where the boundary is rigorously planar and stationary. Any deviation, even slight, from these simplifying hypotheses leads to variable couplings between the two types of discontinuities, the rotational and compressional characters separating and interpenetrating in a complex way. This is observed, for example, in numerical simulations where the tangential initial condition can be perturbed by a "tearing instability", leading to the reconnection of the lines coming from both sides of the current layer.

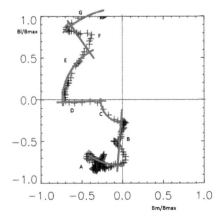

Figure 7.12. *Magnetic field data acquired by Cluster (ESA) on April 6, 2004 at around 4:30 am. The field is projected in the plane tangent to the magnetopause. For a color version of this figure, see www.iste.co.uk/belmont/plasma.zip*

In situ measurements made on board satellites most often show such complex situations. Multi-satellite measurements have indeed shown that the assumption that the boundaries of the magnetosphere are one-dimensional structures that can be considered stationary for the duration of the crossing, is generally not verified. Moreover, rotational and compressional variations sometimes seem to have a tendency to separate.

Figure 7.12 shows an example of a polar representation of the magnetic field in the plane tangent to the magnetopause during a magnetopause crossing observed by a Cluster spacecraft. Starting from a direction of the field in the magnetosheath (around point A), we observe a succession of phases where the field varies in a way that is approximately rotational (A, C, E, G) and phases where it varies in an approximately compressional way (B, D, F), before reaching the direction of the magnetospheric field (around point G).

The environments of other planets are less well explored than the environment of the Earth, but all have a shockwave in the front. However, only planets that have a magnetic field have a magnetosphere and, therefore, a magnetopause. The characteristic dimensions depend on both the distance from the Sun and the intensity of its own magnetic field. There is therefore a great variety: the giant planets have a giant magnetosphere, which is more or less well known.

Figure 7.13 shows the first (and only, to date) measurements of the magnetic field around Uranus when Voyager 2 passed close by (about 3 radii from the planet). We see the presence of a huge magnetosphere that extends over more than 100 radii of the planet. On the contrary, Mercury has a very weak field and a very small magnetosphere (see Figure 7.14). This magnetosphere can sometimes completely disappear.

Figure 7.15 illustrates another aspect of magnetospheric boundaries. It has been mentioned throughout this chapter that these boundaries can be described very well in the context of MHD, but kinetic phenomena also occur. The hybrid simulation of the Earth's shock shows the existence of a pre-shock. Upstream of the shock, ions are oberved that have been reflected on the shock (visible from the disturbances of the magnetic field outside the shock). The geometry of this pre-shock depends on the direction of the interplanetary magnetic field (which is not represented here), and these ions have an influence on the field itself. This pre-shock is also observed experimentally.

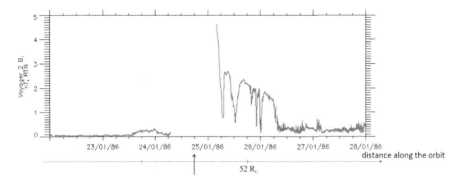

Figure 7.13. *Magnetic field recorded by Voyager 2 as it passed near Uranus in 1986. There is a data gap around the closest approach point to the planet, indicated by the arrow (source: figure made with AMDA). For a color version of this figure, see www.iste.co.uk/belmont/plasma.zip*

Figure 7.14. *MHD modeling of Mercury's magnetosphere. We can see that the size of the magnetosphere on the front side is of the order of the radius of the planet (source: F. Pantellini). For a color version of this figure, see www.iste.co.uk/belmont/plasma.zip*

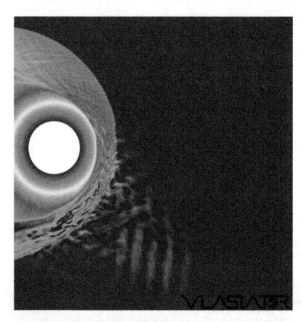

Figure 7.15. *Plot of the magnetic field magnitude in a numerical simulation of the Earth's shock produced with the Vlasiator code (http://www.physics.helsinki.fi/ vlasiator/). It is a hybrid code that describes the ions through the Vlasov equation and the electrons as a fluid. The white circle shows the limit of the simulation, and the Sun is on the right (source: M. Palmroth). For a color version of this figure, see www.iste.co.uk/belmont/plasma.zip*

References

Balescu, R. (1960). Irreversible processes in ionized gases. *Physics of Fluids*, 3(1), 52–63.

Baumjohann, W. and Treumann, R.-A. (1997). *Basic Space Plasma Physics*. Imperial College Press, London.

Belmont, G., Grappin, R., Mottez, F., Pantellini, F., and Pelletier, G. (2013). *Collisionless Plasmas in Astrophysics*. Wiley, Weinheim.

Bhatnagar, P.-L., Gross, E.-P., and Krook, M. (1954). A model for collision processes in gases. I: Small amplitude processes in charged and neutral one-component systems. *Physical Review*, 94(3), 511–525.

Bittencourt, J.-A. (2013). *Fundamentals of Plasma Physics*. Springer, New York.

Braginskii, S.-I. (1965). Transport processes in a plasma. In *Reviews of Plasma Physics 1*, Leontovich, M.A. (ed.). Consultants Bureau, New York.

Chen, F.-F. (1984). *Introduction to Plasma Physics and Controlled Fusion*. Plenum Press, New York.

Chew, G.-F., Goldberger, M.-L., and Low, F.-E. (1956). The Boltzmann equation and the one-fluid hydromagnetic equations in the absence of particle collisions. *Proceedings of the Royal Society of London*, 236(1204), 112–118.

Davidson, R.-C. (1972). *Methods in Non-linear Plasma Theory*. Academic Press, Cambridge.

Davies, K. (1966). *Ionospheric Radio Propagation*. Dover Publications, New York.

de Hoffmann, F. and Teller, R. (1950). Magneto-hydrodynamic shocks. *Physical Review*, 80.

Delcroix, J.-L. and Bers, A. (1994). *Physique des Plasmas*. EDP Sciences, Les Ulis.

Galtier, S. (2013). *Magnétohydrodynamique. Des plasmas de laboratoire à l'astrophysique*. Vuibert, Paris.

Guernsey, R.L. (1962). Kinetic equation for a completely ionized gas. *Physics of Fluids*, 5, 322.

Hull, A.-W. and Langmuir, I. (1929). Control of an arc discharge by means of a grid. *Proceedings of the National Academy of Sciences of the United States of America*, 15(3), 218–225.

Klimontovich, Y.L. (1967). *The Statistical Theory of Non-Equilibrium Process in Plasma*. Elsevier, New York.

Krall, N.-A. and Trivelpiece, A.-W. (1973). *Principles of Plasma Physics*. McGraw-Hill, New York.

Landau, L.D. (1946). On the vibrations of the electronic plasma. *Journal of Physics*, 10, 25–34.

Landau, L.D. (1965). The transport equation in the case of Coulomb interaction. In *Collected Papers of L.D. Landau*, ter Haar, D. (ed.). Pergamon Press, Oxford.

Landau, L.D., Lifchitz, L., and Pitaevskii, L. (1990). *Physique théorique : Tome 8, Électrodynamique des milieux continus*. Éditions MIR, Moscow.

Lenard, A. (1960). On Bogoliubov's kinetic equation for a spatially homogeneous plasma. *Annals of Physics*, 10(3), 390–400.

Quémada, D. (1968). *Ondes dans les plasmas*. Hermann, Paris.

Rax, J.-M. (2005). *Physique des plasmas : Cours et applications*. Dunod, Paris.

Stix, T.-H. (1992). *Waves in Plasmas*. Springer, New York.

van Kampen, N.-G. and Felderhof, B.-U. (1967). *Theoretical Methods in Plasma Physics*. North Holland Publishing Company, Amsterdam.

Index

Printed in the United States
By Bookmasters